Introduction to
Micrometeorology

This is Volume 42 in
INTERNATIONAL GEOPHYSICS SERIES
A series of monographs and textbooks
Edited by RENATA DMOWSKA and JAMES R. HOLTON

A complete list of the books in this series appears at the end of this volume.

Introduction to Micrometeorology

S. Pal Arya

Department of Marine, Earth and
  Atmospheric Sciences
College of Physical and Mathematical Sciences
North Carolina State University
Raleigh, North Carolina

ACADEMIC PRESS, INC.
Harcourt Brace Jovanovich, Publishers
San Diego   New York   Berkeley   Boston
London   Sydney   Tokyo   Toronto

About the Cover: A micrometeorological mast over the Arctic pack ice used during the 1972 Arctic Ice Dynamics Joint Experiment (AIDJEX).

ACADEMIC PRESS, INC.
San Diego, California 92101

United Kingdom Edition published by
ACADEMIC PRESS, INC. (LONDON) LTD.
24-28 Oval Road, London NW1 7DX

Library of Congress Cataloging-in-Publication Data

Arya, S. P. S.
    Introduction to micrometeorology.
    Includes index.
    1. Micrometeorology.   I. Title.
QC883.8.A79   1988   551.5   88-3429
ISBN 0-12-064490-8   (alk. paper)

PRINTED IN THE UNITED STATES OF AMERICA
88 89 90 91   9 8 7 6 5 4 3 2 1

To my wife, Nirmal, and
my children, Niki, Sumi, and Vishal

Contents

Chapter 14   Nonhomogeneous Boundary Layers

Chapter 15   Agricultural and Forest Micrometeorology

References

Preface

Many university departments of meteorology or atmospheric sciences offer a course in micrometeorology as a part of their undergraduate curricula. Some graduate programs also include an introductory course in micrometeorology or environmental fluid mechanics, in addition to more advanced courses on planetary boundary layer (PBL) and turbulence. While there are a number of excellent textbooks and monographs available for advanced graduate courses, there is no suitable text for an introductory undergraduate or graduate course in micrometeorology. My colleagues and I have faced this problem for more than ten years, and we suspect that other instructors of introductory micrometeorology courses have faced the same problem—lack of a suitable textbook. At first, we tried to use R. E. Munn's "Descriptive Micrometeorology," but soon found it to be much out-of-date. There has been almost an explosive development of the field during the past two decades, as a result of increased interest in micrometeorological problems and advances in computational and observational facilities. When Academic Press approached me to write a textbook, I immediately saw an opportunity for filling a void and satisfying an acutely felt need for an introductory text, particularly for undergraduate students and instructors. I also had in mind some incoming graduate students having no background in micrometeorology or fluid mechanics. Instructors may also find the latter half of the book suitable for an introductory or first graduate course in atmospheric boundary layer and turbulence. Finally, it may serve as an information source for boundary-layer meteorologists, air pollution meteorologists, agricultural and forest micrometeorologists, and environmental scientists and engineers.

Keeping in mind my primary readership, I have tried to introduce the various topics at a sufficiently elementary level, starting from the basic thermodynamic and fluid dynamic laws and concepts. I have also given qualitative descriptions based on observations before introducing more complex theoretical concepts and quantitative relations. Mathematical treatment is deliberately kept simple, presuming only a minimal mathematical background of upper-division science majors. Uniform notation and symbols are used throughout the text, although certain symbols do represent different things in different contexts. The list of symbols should be helpful to the reader. Sample problems and exercises given at the end of each chapter should be useful to students, as well as to instructors. Many of these were given in homework assignments, tests, and examinations for our undergraduate course in micrometeorology.

The book is organized in the form of fifteen chapters, arranged in order of increasing complexity and what I considered to be a natural order of the topics covered in the book. The scope and importance of micrometeorology and turbulent exchange processes in the PBL are introduced in Chapter 1. The next three chapters describe the energy budget near the surface and its components, such as radiative, conductive, and convective heat fluxes. Chapter 5 reviews basic thermodynamic relations and presents typical temperature and humidity distributions in the PBL. Wind distribution in the PBL and simple dynamics, including the balance of forces on an air parcel in the PBL, are discussed in Chapter 6. The emphasis, thus far, is on observations, and very little theory is used in these early chapters. The viscous flow theory, fundamentals of turbulence, and classical semiempirical theories of turbulence, which are widely used in micrometeorology and fluid mechanics, are introduced in Chapters 7, 8, and 9. Instructors of undergraduate courses may like to skip some of the material presented here, particularly the Reynolds-averaged equations for turbulent motion. Chapters 10–12 present the surface-layer similarity theory and micrometeorological methods and observations using the similarity framework. These three chapters would constitute the core of any course in micrometeorology. The last three chapters cover the more specialized topics of the marine atmospheric boundary layer, nonhomogeneous boundary layers, and micrometeorology of vegetated surfaces. Since these topics are still hotly pursued by researchers, the reader will find much new information taken from recent journal articles. The instructor should be selective in what might be included in an undergraduate course. Chapters 7–15 should also be suitable for a first graduate course in micrometeorology or PBL.

Finally, I would like to acknowledge the contribution of my colleagues, Jerry Davis, Al Riordan, and Sethu Raman, and of my students for reviewing certain parts of the manuscript and pointing out many errors and omissions. I am most grateful to Professor Robert Fleagle for his thorough and timely review of each chapter, without the benefit of the figures and illustrations. His comments and suggestions were extremely helpful in preparing the final draft. I would also like to thank the publishers and authors of many journal articles, books, and monographs for their permission to reproduce many figures and tables in this book. The original sources are acknowledged in figure legends, whenever applicable. Still, many ideas, explanations, and discussions in the text are presented without specific acknowledgments of the original sources, because, in many cases, the original sources were not known and also because, I thought, giving too many references in the text would actually distract the reader. In the end, I wish to acknowledge the help of Brenda Batts and Dava House for typing the manuscript and Pat Bowers and LuAnn Salzillo for drafting figures and illustrations.

Symbols

A Amplitude of thermal wave in the subsurface medium (Chapter 4)

A Aerodynamic surface area of vegetation per unit volume (Chapter 15)

A_s Amplitude of thermal wave at the surface

A_θ Mean advection terms in the thermodynamic energy equation

a Albedo or shortwave reflectivity of the surface (Chapter 3)

a $a \equiv (f/2\nu)^{1/2}$ (Chapter 7)

a An empirical constant (Chapter 13)

a_b Acceleration due to buoyancy

a_i Empirical constant

a_θ Instantaneous advection terms in the thermodynamic energy equation

a_1 Empirical constant

a_2 Empirical constant

B Bowen ratio

B Buoyancy production term in the turbulent kinetic energy equation (Chapter 8)

b Boltzmann constant

C Heat capacity

C A constant (Chapter 9)

C_D Surface drag coefficient

C_d Drag coefficient (Chapter 15)

C_{DN} Drag coefficient in neutral stability

C_H Heat transfer coefficient

C_{HN} Heat transfer coefficient in neutral stability

C_W Water vapor transfer coefficient

C_w Wave speed

C_w A constant (Chapter 9)

C_θ An empirical constant

C_0 Phase speed of dominant waves

c Specific heat

c Speed of light (electromagnetic waves) in vacuum (Chapter 3)

c_m Electromagnetic wave speed in a medium (Chapter 3)

c_m A constant (Chapter 14)

c_p Specific heat at constant pressure

c_v Specific heat at constant volume

D Reference depth in the submedium (Chapter 4)

D Viscous dissipation term in the turbulent kinetic energy equation (Chapter 8)

D/Dt Total derivative

d Distance of the earth from the sun (Chapter 3)

d Damping depth of thermal wave in the submedium (Chapter 4)

dH Heat added to a parcel

d_m Mean distance between the earth and the sun

dP Change in the air pressure

dT Change in the air temperature

d_0 Zero-plane displacement

E Rate of evaporation or condensation

E_a Drying power of air

E_p Potential evaporation or evapotranspiration

E_0 Rate of evaporation or condensation at the surface

e Water vapor pressure

e_s Water vapor pressure at saturation

e_0 Water vapor pressure at the surface or at $z = z_0$

F Function of certain variables (Chapter 9)

F Froude number (Chapters 13 and 14)

F_c Critical Froude number

F_L Froude number based on length of the hill

F_u A similarity function

F_v A similarity function

f Coriolis parameter

f Function of variables (Chapter 9)

f' Normalized velocity as a function of η

G Magnitude of geostrophic wind

$\mathbf{G}$ Geostrophic wind vector

Gr Grashof number

G_0 Magnitude of surface geostrophic wind

$\mathbf{G}_0$ Surface geostrophic wind vector

g Acceleration due to gravity

H Sensible (direct) heat flux

H Wave height (Chapter 13)

H Building height (Chapter 14)

$\overline{H}$ Mean wave height

H_a Anthropogenic heat flux in an urban area

H_G Ground heat flux to or from the subsurface medium

H_i Heat flux at the inversion base

H_{in} Energy coming in

H_L Latent heat flux

H_{out} Energy going out

H_s Height of the dividing streamline

H_0 Sensible heat flux at the surface

$\overline{H}_{1/3}$ Significant wave height

h Boundary layer thickness

h Depth of channel flow or distance between two parallel planes (Chapter 7)

h_E Ekman depth

h_i Internal boundary layer thickness

h_p Planck's constant

h_0 Average height of roughness elements or plant canopy

I Insolation at the surface

I_0 Insolation at the top of the atmosphere

i Imaginary number $\sqrt{-1}$

K Effective viscosity

K_h Eddy diffusivity of heat

K_m Eddy viscosity or diffusivity of momentum

K_{mr} Eddy viscosity at the reference height

K_w Eddy diffusivity of water vapor

k von Karman constant

k Thermal conductivity (Chapter 4)

k_m Hydraulic conductivity of the subsurface medium

L Obukhov length

L A characteristic length scale (Chapter 9)

L Wavelength (Chapter 13)

L_{AI} Leaf area index

L_e Latent heat of evaporation/ condensation

l Fluctuating mixing length

l_m Mean mixing length

M Soil moisture flow rate

M_b Soil moisture flow rate at the bottom of layer

M_d Mass of dry air

M_w Mass of water vapor

M_0 Soil moisture flow rate at the surface

m Mean molecular mass of air (Chapter 5)

m Exponent in the power-law wind profile equation (Chapter 10)

m A coefficient (Chapter 15)

m_d Mean molecular mass of dry air

m_w Mean molecular mass of water vapor

N Brunt–Vaisala frequency

Nu   Nusselt number

n Exponent in the power-law eddy viscosity profile (Chapter 10)

n Wave frequency (Chapter 13)

n A coefficient (Chapter 15)

P Mean air pressure

P Period of the wave (Chapter 4)

P_{AI} Plant area index

Pe   Peclet number

Pr   Prandtl number

P_0 Pressure of the reference atmosphere

p Instantaneous pressure

p_0 Pressure of the reference state

p_1 Deviation of pressure from the reference state

Q Mean specific humidity

Q_s Mean specific humidity at saturation

Q_0 Specific humidity at $z = z_0$

q Specific humidity fluctuation

q_* Specific humidity scale for the surface layer

R Radiative energy flux or radiation (Chapter 3)

R Specific gas constant (Chapter 5)

Ra   Rayleigh number

R_D Diffuse radiation

R_d Specific gas constant for dry air

Re   Reynolds number

Ri   Richardson number

Ri_B Bulk Richardson number

Ri_c Critical Richardson number

R_L Longwave radiation (flux)

$R_{L\downarrow}$ Incoming (downward) longwave radiation

$R_{L\uparrow}$ Outgoing (upward) longwave radiation

R_N Net radiation

R_S Shortwave radiation (flux)

$R_{S\downarrow}$ Incoming (downward) shortwave radiation

$R_{S\uparrow}$ Outgoing (upward) shortwave radiation

R_s Solar radiation flux at the surface

R_* Absolute gas constant

R_λ Radiative energy flux density per unit wavelength

R_0 Solar radiation flux at the top of the atmosphere

r_H Aerial resistance to heat transfer

r_M Resistance to momentum transfer

r_W Resistance to water vapor transfer

S Shear production term in the turbulent kinetic energy equation (Chapter 8)

S Soil moisture content (Chapter 12)

S_F Speed-up factor

S_0 Solar constant

s Static stability

T Mean air temperature

T Temperature of the subsurface medium (Chapter 4)

T Wave period (Chapter 13)

T_{eb} Equivalent black body temperature

T_m Mean temperature of the surface and submedium

T_r	Turbulent transport of turbulent kinetic energy
T_s	Surface temperature (Chapter 4)
T_s	Sea-surface temperature (Chapter 13)
T_u	Urban air temperature
T_v	Mean virtual temperature of moist air
T_{vp}	Mean virtual temperature of the parcel
T_{v0}	Virtual temperature at the reference state
T_*	Convective temperature scale
T_0	Air temperature at the reference state
T_1	Deviation in the temperature from the reference state
T_{10}	Air temperature at 10 m above the surface
t	Time
U	Mean velocity component in x direction
U_g	Geostrophic wind component in x direction
U_h	Velocity at $z = h$
U_h	Velocity of the moving surface (Chapter 7)
U_m	Mixed-layer averaged wind speed
U_r	Wind speed at the reference height
U_∞	Ambient wind speed
U_0	Mean velocity in the approach flow

U_{10}	Wind speed at a reference height of 10 m
u	Fluctuating velocity component in x direction
u	Characteristic velocity scale of turbulence (Chapter 8)
$\bar{u}$	Instantaneous velocity in x direction
u_f	Local free convective velocity scale
u_*	Friction velocity
V	Mean velocity component in y direction
V	Volume (Chapter 2)
V	Characteristic velocity scale (Chapter 9)
$\mathbf{V}$	Wind vector
V_g	Geostrophic wind component in y direction
v	Fluctuating velocity component in y direction
$\bar{v}$	Instantaneous velocity in y direction
W	Mean velocity component in z (vertical) direction
W	Building width (Chapter 14)
W_{AI}	Woody-element area index
W_*	Convective velocity scale
w	Fluctuating velocity component in z (vertical) direction
$\bar{w}$	Instantaneous velocity in z direction
z	Height above or depth below the surface or an appropriate reference plane

z_m Geometric mean of the two heights z_1 and z_2

z_r Reference height

z' Height above the ground level

z_0 Roughness length parameter

z_{01} Roughness parameter of the upstream surface

z_{02} Roughness parameter of the downstream surface

α Cross-isobar angle (Chapter 6)

α Empirical constant (Chapter 15)

α_h Molecular diffusivity of heat or thermal diffusivity

α_m Soil moisture diffusivity

α_w Molecular diffusivity of water vapor in air

α_λ Absorptivity at wavelength λ

α_0 Cross-isobar angle at the surface

β Coefficient of thermal expansion

β Slope angle of an inclined plane (Chapter 7)

Γ Adiabatic lapse rate

Γ_s Saturated adiabatic lapse rate

γ Solar zenith angle (Chapter 3)

γ Psychrometer constant (Chapter 12)

Δ Gradient operator

Δ Slope of the saturation vapor pressure versus temperature curve (Chapter 12)

ΔA Elemental area

ΔH_S Rate of energy storage

ΔM Rate of storage of soil moisture

ΔT_s Difference between the maximum and minimum sea surface temperatures

ΔT_{u-r} Difference between the urban and rural air temperatures

ΔU Difference in mean velocities at two heights

ΔU Velocity deficit in the wake

ΔV_g Change in V_g over a layer

$\Delta \Theta$ Difference between potential temperatures at two heights

Δx Small increment in x

Δy Small increment in y

Δz Small increment in z

Δz $z_2 - z_1$

∇_2 Laplacian operator

ε Overall infrared emissivity

ε_λ Emissivity at wavelength λ

ζ Monin–Obukhov stability parameter

η Normalized distance from the surface (Chapter 7)

η Sea-surface elevation (Chapter 13)

Θ Mean potential temperature

Θ_m Mixed-layer averaged potential temperature

Θ_r Rural air potential temperature

Θ_u Urban air potential temperature

Θ_v Mean virtual potential temperature

Θ_0 Potential temperature at $z = z_0$

Θ_{01} Temperature of the upstream surface

Θ_{02} Temperature of the downstream surface

θ Fluctuating potential temperature

$\tilde{\theta}$ Instantaneous potential temperature

θ_f Local free convective temperature scale

κ Wave number

λ Wavelength

λ_{max} Wavelength corresponding to spectral maximum

μ Dynamic viscosity

ν Kinematic viscosity

Π Dimensionless group

π Ratio of circumference to diameter of a circle

ρ Mass density of air or other medium

ρ_p Mass density of the air parcel

ρ_0 Density of the reference state

ρ_1 Deviation in the density from the reference state

σ Stefan–Boltzmann constant

σ_q Standard deviation of specific humidity fluctuations

σ_u Standard deviation of velocity fluctuations in x direction

σ_v Standard deviation of velocity fluctuations in y direction

σ_w Standard deviation of velocity fluctuations in z direction

σ_η Standard deviation of sea-surface elevation

σ_θ Standard deviation of temperature fluctuations

τ Shearing stress

$\boldsymbol{\tau}$ Shear stress vector

τ_0 Surface shear stress or drag

ϕ_h Dimensionless potential temperature gradient

ϕ_m Dimensionless wind shear

ϕ_w Dimensionless specific humidity gradient

ψ_h M–O similarity function for normalized potential temperature

ψ_m M–O similarity function for normalized velocity

ψ_w M–O similarity function for normalized specific humidity

Ω Rotational speed of the earth

$\boldsymbol{\Omega}$ Earth's rotational velocity vector

ω Wave frequency

Chapter 1 | Introduction

1.1 SCOPE OF MICROMETEOROLOGY

Atmospheric motions are characterized by a variety of scales ranging from the order of a millimeter to as large as the circumference of the earth in the horizontal direction and the entire depth of the atmosphere in the vertical direction. The corresponding time scales range from a tiny fraction of a second to several months or years. These scales of motions are generally classified into three broad categories, namely, micro-, meso-, and macroscales. Sometimes, terms such as local, regional, and global are used to characterize the atmospheric scales and the phenomena associated with them.

Micrometeorology is a branch of meteorology which deals with the atmospheric phenomena and processes at the lower end of the spectrum of atmospheric scales, which are variously characterized as microscale, small-scale, or local-scale processes. The scope of micrometeorology is further limited to only those phenomena which originate in and are dominated by the shallow layer of frictional influence adjoining the earth's surface, commonly known as the atmospheric boundary layer (ABL) or the planetary boundary layer (PBL). Thus, some of the small-scale phenomena, such as convective clouds and tornados, are considered outside the scope of micrometeorology, because their dynamics is largely governed by mesoscale and macroscale weather systems.

1.1.1 ATMOSPHERIC BOUNDARY LAYER

A boundary layer is defined as the layer of a fluid (liquid or gas) in the immediate vicinity of a material surface in which significant exchange of momentum, heat, or mass takes place between the surface and the fluid. Sharp variations in the properties of the flow, such as velocity, temperature, and mass concentration, also occur in the boundary layer.

The atmospheric boundary layer is formed as a consequence of the

1

interactions between the atmosphere and the underlying surface (land or water) over time scales of a few hours to about 1 day. Over longer periods the earth–atmosphere interactions may span the whole depth of the troposphere, typically 10 km, although the PBL still plays an important part in these interactions. The influence of surface friction, heating, etc., is quickly and efficiently transmitted to the entire PBL through the mechanism of turbulent transfer or mixing. Momentum, heat, and mass can also be transferred downward through the PBL to the surface through the same mechanism.

The atmospheric PBL height varies over a wide range (several tens of meters to several kilometers) and depends on the rate of heating or cooling of the surface, strength of winds, the roughness and topographical characteristics of the surface, large-scale vertical motion, horizontal advections of heat and moisture, and other factors. In the air pollution literature the PBL height is commonly referred to as the mixing depth, since it represents the depth of the layer through which pollutants released from the surface and in the PBL get eventually mixed. As a result, the PBL is dirtier than the free atmosphere above it. The contrast between the two is usually quite sharp over large cities and can be observed from an aircraft as it leaves or enters the PBL.

Following sunrise on a clear day, the continuous heating of the surface by the sun and the resulting thermal mixing in the PBL cause the PBL depth to increase steadily throughout the day and attain a maximum value of the order of 1 km (range $\simeq$ 0.2–5 km) in the late afternoon. Later in the evening and throughout the night, on the other hand, the radiative cooling of the ground surface results in the suppression or weakening of turbulent mixing and consequently in the shrinking of the PBL depth to a typical value of the order of only 100 m (range $\simeq$ 20–500 m). Thus the PBL depth waxes and wanes in response to the diurnal heating and cooling cycle. The winds, temperatures, and other properties of the PBL may also be expected to exhibit strong diurnal variations.

1.1.2 THE SURFACE LAYER

Some investigators limit the scope of micrometeorology to only the so-called atmospheric surface layer, which comprises the lowest one-tenth or so of the PBL and in which the earth's rotational or Coriolis effects can be ignored. Such a restriction may not be desirable, because the surface layer is an integral part of and is much influenced by the PBL as a whole, and the top of this layer is not physically as well defined as the top of the PBL. The latter represents the fairly sharp boundary between the irregular and almost chaotic (turbulent) motions in the PBL and the considerably smooth and streamlined (nonturbulent) flow in the free atmosphere above. The PBL top can be easily detected by ground-based remote-sensing devices, such as acoustic sounder, lidar, etc., and can also be inferred from temperature, humidity, and wind soundings.

However, the surface layer is more readily amenable to observations from the surface, as well as from micrometeorological masts and towers. It is also the layer in which most human beings, animals, and vegetation live and in which most human activities take place. The sharpest variations in meteorological variables with height occur within the surface layer and, consequently, the most significant exchanges of momentum, heat, and mass also occur in this layer. Therefore, it is not surprising that the surface layer has received far greater attention from micrometeorologists and microclimatologists than has the outer part of the PBL.

1.1.3   TURBULENCE

Turbulence refers to the apparently chaotic nature of many flows, which is manifested in the form of irregular, almost random fluctuations in velocity, temperature, and scalar concentrations around their mean values in time and space. The motions in the atmospheric boundary layer are almost always turbulent. In the surface layer, turbulence is more or less continuous, while it may be intermittent and patchy in the upper part of the PBL and is sometimes mixed with internal gravity waves.

Near the surface, atmospheric turbulence manifests itself through the flutter of leaves of trees and blades of grass, swaying of branches of trees and plants, irregular movements of smoke and dust particles, generation of ripples and waves on water surfaces, and variety of other visible phenomena. In the upper part of the PBL, turbulence is manifested by irregular motions of kites and balloons, spreading of smoke and other visible pollutants as they exit tall stacks or chimneys, and fluctuations in the temperature and refractive index encountered in the transmission of sound, light, and radio waves.

1.2   MICROMETEOROLOGY VERSUS MICROCLIMATOLOGY

The difference between micrometeorology and microclimatology is primarily in the time of averaging the variables, such as air velocity, temperature, humidity, etc. While the micrometeorologist is primarily interested in fluctuations, as well as in the short-term (of the order of an hour or less) averages of meteorological variables in the PBL or the surface layer, the microclimatologist mainly deals with the long-term (climatological) averages of the same variables. The latter is also interested in diurnal and seasonal variations, as well as in very long-term trends in meteorological parameters.

In micrometeorology, detailed examination of small-scale temporal and spatial structure of flow and thermodynamic variables is often necessary to gain a better understanding of the phenomena of interest. For example, in order to determine the short-time average concentrations on the ground some distance away from a pollutant source, one needs to know not only the mean winds that are responsible for the mean transport of the pollutant, but also the statistical properties of wind gusts (turbulent fluctuations) that are responsible for spreading or diffusing the pollutant as it moves along the mean wind.

In microclimatology, one deals with long-term averages of meteorological variables in the near-surface layer of the atmosphere. Details of fine structure are not considered so important, because their effects on the mean variables are smoothed out in the averaging process. Still, one cannot ignore the long-term consequences of the small-scale turbulent transfer and exchange processes.

Despite the above-mentioned differences, micrometeorology and microclimatology have much in common, because they both deal with similar atmospheric processes occurring near the surface. Their interrelationship is further emphasized by the fact that long-term averages dealt with in the latter can, in principle, be obtained by the integration in time of the short-time averaged micrometeorological variables. It is not surprising, therefore, to find some of the fundamentals of micrometeorology described in books on microclimatology (e.g., Geiger, 1965; Oke, 1978; Rosenberg *et al.*, 1983). Likewise, microclimatological information is found in and serves useful purpose in texts on micrometeorology (e.g., Sutton, 1953; Munn, 1966).

1.3 IMPORTANCE AND APPLICATIONS OF MICROMETEOROLOGY

Although the atmospheric boundary layer comprises only a tiny fraction of the atmosphere, the small-scale processes occurring within the PBL are useful to various human activities and are important for the well being and even survival of life on earth. This is not merely because the air near the ground provides the necessary oxygen to human beings and animals, but also because this air is always in turbulent motion, which causes efficient mixing of pollutants and exchanges of heat, water vapor, etc., with the surface.

1.3.1 TURBULENT TRANSFER PROCESSES

Turbulence is responsible for the efficient mixing and exchange of mass, heat, and momentum throughout the PBL. In particular, the surface layer turbulence is responsible for exchanging these properties between the atmosphere and the earth's surface. Without turbulence, such exchanges would have been at the molecular scale and miniscule in magnitudes (10^{-3}–10^{-6} times the turbulent transfers that now commonly occur). Nearly all the energy which drives the large-scale weather and general circulation comes through the PBL.

Through the efficient transfer of heat and moisture, the boundary layer turbulence moderates the microclimate near the ground and makes it habitable for animals, organisms, and plants. The atmosphere receives virtually all of its water vapor through turbulent exchanges near the surface. Evaporation from land and water surfaces is not only important in the surface water budget and the hydrological cycle, but the latent heat of evaporation is also an important component of the surface energy budget. This water vapor, when condensed on tiny dust particles and other aerosols (cloud condensation nuclei), leads to the formation of fog, haze, and clouds in the atmosphere.

Besides water vapor, there are other important exchanges of mass within the PBL involving a variety of gases and particulates. Turbulence is important in the exchange of carbon dioxide between plants and animals. Through efficient diffusion of the various pollutants released near the ground and mixing them throughout the PBL and parts of the lower troposphere, atmospheric turbulence prevents the fatal poisoning of life on earth. The quality of air we breathe depends to a large extent on the mixing capability of the PBL turbulence. The boundary layer turbulence also picks up pollen and other seeds of life, spreads them out, and deposits them over wide areas far removed from their origin. It lifts dust, salt particles, and other aerosols from the surface and spreads them throughout the lower atmosphere. Of these, the so-called cloud condensation nuclei are an essential ingredient in the condensation and precipitation processes in the atmosphere.

Through the above-mentioned mass exchange processes between the earth's surface and the atmosphere, the radiation balance and the heat energy budget at or near the earth's surface are also significantly affected. More direct effects of turbulent transfer on the surface heat energy budget are through sensible and latent heat exchanges between the surface and the atmosphere. Over land, the sensible heat exchange is usually more important than the latent heat of evaporation, but the reverse is true over large lakes and the oceans.

Turbulent transfer of momentum between the earth and atmosphere is also very important. It is essentially a one-way process in which the earth acts as a sink of atmospheric momentum (relative to earth). In other words, the earth's surface exerts frictional resistance to atmospheric motions and slows them down in the process. The moving air near the surface may be thought of exerting an equivalent drag force on the surface. The rougher the surface, the larger would be the drag force per unit area of the surface. Some commonly encountered examples of this are the marked slowdown of surface winds in going from a large lake, bay, or sea to inland areas and from rural to urban areas. Perhaps the most vivid manifestations of the effect of increased surface drag are the rapid weakening and subsequent demise of hurricanes and other tropical storms as they move inland. Another factor responsible for this phenomenon is the marked reduction or cutoff of the available latent heat energy to the storm from the surface.

Over large lakes and oceans, wind drag is responsible for the generation of waves and currents in water, as well as for the movement of sea ice. In coastal areas, wind drag causes storm surges and beach erosion. Over land, drag exerts wind loads on vehicles, buildings, towers, bridges, cables, and other structures. Wind stress over sand surfaces raises dust and creates ripples and sand dunes. When accompanied by strong winds, the surface layer turbulence can be quite discomforting and even harmful to people, animals, and vegetation.

Turbulent exchange processes in the PBL have profound effects on the evolution of local weather. Boundary layer friction is primarily responsible for the low-level convergence and divergence of flow in the regions of lows and highs in surface pressures, respectively. The frictional convergence in a moist boundary layer is also responsible for the low-level convergence of moisture in low-pressure regions. The kinetic energy of the atmosphere is continuously dissipated by small-scale turbulence in the atmosphere. Almost one-half of this loss on an annual basis occurs within the PBL, even though the PBL comprises only a tiny fraction (less than 2%) of the total kinetic energy of the atmosphere.

1.3.2 APPLICATIONS

In the following text, we have listed some of the possible areas of application of micrometeorology together with the subareas or activities in which micrometeorological information may be especially useful.

1. Air pollution meteorology
   - Atmospheric transport and diffusion of pollutants
   - Atmospheric deposition on land and water surfaces

- Prediction of local and regional air quality
- Selection of sites for power plants and other major sources of pollution
- Selection of sites for monitoring urban and regional air pollution
- Industrial operations
- Agricultural operations, such as dusting, spraying, and burning
- Military operations
2. Mesoscale meteorology
    - Urban boundary layer and heat island
    - Land–sea breezes
    - Drainage and mountain valley winds
    - Dust devils, water spouts, and tornados
    - Development of fronts and cyclones
3. Macrometeorology
    - Atmospheric predictability
    - Long-range weather forecasting
    - Siting and exposure of meteorological stations
    - General circulation and climate modeling
4. Agricultural and forest meteorology
    - Prediction of surface temperatures and frost conditions
    - Soil temperature and moisture
    - Evapotranspiration
    - Radiation balance of a plant cover
    - Carbon dioxide exchanges within the plant canopy
    - Temperature, humidity, and winds in the canopy
    - Protection of crops and shrubs from strong winds and frost
    - Wind erosion of soil and protective measures
    - Effects of acid rain and other pollutants on plants and trees
5. Urban planning and management
    - Prediction and abatement of ground fogs
    - Heating and cooling requirements
    - Wind loading and designing of structures
    - Wind sheltering and protective measures
    - Instituting air pollution control measures
6. Physical oceanography
    - Prediction of storm surges
    - Prediction of the sea state
    - Dynamics of the oceanic mixed layer
    - Movement of sea ice
    - Modeling large-scale oceanic circulations
    - Navigation
    - Radio transmission

PROBLEMS AND EXERCISES

1. Compare and contrast the following terms:
   (a) Microscale and macroscale atmospheric processes and phenomena
   (b) Micrometeorology and microclimatology
   (c) Atmospheric boundary layer and the surface layer
2. On a schematic of the lower atmosphere, including the tropopause, indicate the PBL and the surface layer during typical fair-weather, daytime conditions.
3. Discuss the importance of turbulent exchange processes between the earth and the atmosphere.

Chapter 2 | Energy Budget near the Surface

2.1 ENERGY FLUXES AT AN IDEAL SURFACE

The flux of a property in a given direction is defined as its amount per unit time passing through a unit area normal to that direction. In this chapter, we will be concerned with fluxes of the various forms of heat energy at or near the surface. The SI units of energy flux are $J \, sec^{-1} \, m^{-2}$ or $W \, m^{-2}$.

The "ideal" surface considered here is relatively smooth, horizontal, homogeneous, extensive, and opaque to radiation. The energy budget of such a surface is considerably simplified in that only the vertical fluxes of energy need to be considered.

There are essentially four types of energy fluxes at an ideal surface, namely, the net radiation to or from the surface, the sensible (direct) and latent (indirect) heat fluxes to or from the atmosphere, and the heat flux into or out of the submedium (soil or water). The net radiative flux is a result of the radiation balance at the surface, which will be discussed in more detail in Chapter 3. During the daytime, it is usually dominated by the solar radiation and is almost always directed toward the surface, while at night the net radiation is much weaker and directed away from the surface. As a result, the surface warms up during the daytime, while it cools during the evening and night hours, especially under clear sky and undisturbed weather conditions.

The direct or sensible heat flux at and above the surface arises as a result of the difference in the temperatures of the surface and the air above. Actually, the temperature in the atmospheric surface layer varies continuously with height, with the magnitude of the vertical temperature gradient usually decreasing with height. In the immediate vicinity of the interface, the primary mode of heat transfer in air is conduction through molecular exchanges. At distances beyond a few millimeters (the thickness of the so-called molecular sublayer) from the interface, however, the

9

primary mode of heat exchange becomes convection through mean and turbulent motions in the air. The sensible heat flux is usually directed away from the surface during the daytime hours, when the surface is warmer than the air above, and vice versa during the evening and night-time periods.

The latent heat or water vapor flux is a result of evaporation, evapo-transpiration, or condensation at the surface and is given by the product of the latent heat of evaporation or condensation and the rate of evaporation or condensation. Evaporation occurs from water surfaces as well as from moist soil and vegetative surfaces, whenever the air above is drier, i.e., it has lower specific humidity than the air in the immediate vicinity of the surface and its transpiring elements (e.g., leaves). This is usually the situation during the daytime. On the other hand, condensation in the form of dew may occur on relatively colder surfaces at nighttime. The water vapor transfer through air does not involve any real heat exchange, except where phase changes between liquid water and vapor actually take place. Nevertheless, evaporation results in some cooling of the surface which in the surface energy budget is represented by the latent heat flux from the surface to the air above. The ratio of the sensible heat flux to the latent heat flux is called the Bowen ratio.

The heat exchange through the ground medium is primarily due to conduction if the medium is soil, rock, or concrete. Through water, how-ever, heat is transferred in the same way as it is through air, first by conduction in the top few millimeters (molecular sublayer) from the sur-face and then by convection in the deeper layers (surface layer, mixed layer, etc.) of water in motion. The depth of the submedium, which re-sponds to and is affected by changes in the energy fluxes at the surface on a diurnal basis, is typically less than a meter for land surfaces and several tens of meters for large lakes and oceans.

2.2   ENERGY BALANCE EQUATIONS

2.2.1   THE SURFACE ENERGY BUDGET

For deriving a simplified equation for the energy balance at an ideal surface, we assume the surface to be a very thin interface between the two media (air and soil or water), having no mass and heat capacity. The energy fluxes would flow in and out of such a surface without any loss or gain due to the surface. Then, the principle of the conservation of energy at the surface can be expressed as

$$R_N = H + H_L + H_G \qquad (2.1)$$

where R_N is the net radiation, H and H_L are the sensible and latent heat fluxes to or from the air, and H_G is the ground heat flux to or from the submedium. Here we use the sign convention that all the radiative fluxes directed toward the surface are positive, while other (nonradiative) energy fluxes directed away from the surface are positive and vice versa. The Bowen ratio is defined as $B \equiv H/H_L$ and may vary over a very wide range. The latent heat flux can be expressed as $H_L = L_e E$, where $L_e \simeq 2.45 \times 10^6$ J kg^{-1} is the latent heat of evaporation/condensation and E is the rate of evaporation/condensation.

Equation (2.1) describes how the net radiation at the surface must be balanced by a combination of the sensible and latent heat fluxes to the air and the heat flux to the subsurface medium. During the daytime, the surface receives radiative energy ($R_N > 0$), so that it must give out heat to one or both of the media. Typically, H, H_L, and H_G are all positive over land surfaces during the day. This situation is schematically shown in Fig. 2.1a. Actual magnitudes of the various components of the surface energy budget depend on many factors, such as the type of surface and its characteristics (soil moisture, texture, vegetation, etc.), geographical location, month or season, time of day, and weather. Under special circumstances, e.g., when irrigating a field, H and/or H_G may become negative

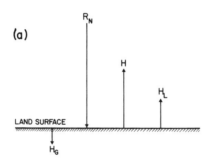

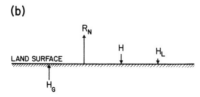

Fig. 2.1 Schematic representation of typical surface energy budgets during (a) daytime and (b) nighttime.

and the latent heat flux due to evaporative cooling of the surface may exceed the net radiation received at the surface.

At night, the surface loses energy by outgoing radiation, especially during clear or partially overcast conditions. This loss is compensated by gains of heat from air and soil media and, at times, from the latent heat of condensation released during the process of dew formation. Thus, according to our sign convention, all the terms of the surface energy balance [Eq. (2.1)] for land surfaces are usually negative during the evening and nighttime periods. Their magnitudes are generally much smaller than the magnitudes of the daytime fluxes, except for H_G. The magnitudes of H_G do not differ widely between day and night, although the direction or sign obviously reverses during the morning and evening transition periods, when other fluxes are also changing their signs (this does not happen simultaneously for all the fluxes, however). Figure 2.1b gives a schematic representation of the surface energy balance during the nighttime.

The energy budgets of extensive water surfaces (large lakes, seas, and oceans) differ from those of land surfaces in several important ways. In the former, the combined value of H_L and H_G balances most of the net radiation, while H plays only a minor role ($H \ll H_L$, or $B \ll 1$). Since the water surface temperature does not respond readily to solar heating due to the large heat capacity and depth of the subsurface mixed layer of a large lake or ocean, the air–water exchanges (H and H_L) do not undergo large diurnal variations.

An important factor to be considered in the energy balance over water surfaces is the penetration of solar radiation to depths of tens of meters. Radiation processes occurring within large bodies of water are not well understood. The radiative fluxes on both sides of the air–water interface must be measured in order to determine the net radiation at the surface. This is not easy to do in the field. Therefore, the surface energy budget as expressed by Eq. (2.1) may not be very useful or even appropriate to consider over water surfaces. A better alternative is to consider the energy budget of the whole energy-active water layer.

2.2.2 ENERGY BUDGET OF A LAYER

An "ideal," horizontally homogeneous, plane surface, which is also opaque to radiation, is rarely encountered in practice. More often, the earth's surface has horizontal inhomogeneities at small scale (e.g., plants, trees, houses, and building blocks), mesoscale (e.g., urban–rural differences, coastlines, hills, and valleys), and large scale (e.g., large mountain ranges). It may be partially transparent to radiation (e.g., water, tall grass, and crops). The surface may be sloping or undulating. In many practical situations, it will be more appropriate to consider the energy budget of a finite interfacial layer, which may include the small-scale surface inhomogeneities and/or the upper part of the subsurface medium which might be active in radiative exchanges. This layer must have finite mass and heat capacity which would allow the energy to be stored in or released from the layer over a given time interval. Such changes in the energy storage with time must, then, be considered in the energy budget for the layer.

If the surface is relatively flat and homogeneous, so that the interfacial layer can be considered to be bounded by horizontal planes at the top and the bottom, one can still use a simplified one-dimensional energy budget for the layer:

$$R_N = H + H_L + H_G + \Delta H_S \qquad (2.2)$$

where ΔH_S is the change in the energy storage per unit time, per unit area in the layer. Thus, the main difference in the energy budget of an interfacial layer from that of an ideal surface is the presence of the rate of energy storage term ΔH_S in the former. Strictly speaking, the various energy fluxes in Eq. (2.2), except for ΔH_S, are net vertical fluxes going in or out of the top and bottom faces of the layer. In reality, however, H and H_L are associated only with the top surface, H_G with the bottom surface, and R_N generally only with the top surface (for an interfacial layer involving the water medium, the bottom boundary may be chosen so that there is no significant radiative flux in or out of the bottom surface), as shown schematically in Fig. 2.2a.

The rate of heat energy storage in the layer can be expressed as

$$\Delta H_S = \int \frac{\partial}{\partial t} (\rho c T) \, dz \qquad (2.3)$$

in which ρ is the mass density, c is the specific heat, T is the absolute temperature of the material at some level z, and the integral is over the whole depth of the layer. When the heat capacity of the medium can be assumed to be constant, independent of z, Eq. (2.3) gives a direct relationship between the rate of energy storage and the rate of warming or cooling of the layer.

The energy storage term ΔH_S in Eq. (2.2) may also be interpreted as the difference between the energy coming in (H_{in}) and the energy going out (H_{out}) of the layer, where H_{in} and H_{out} represent appropriate combinations of R_N, H, H_L, and H_G, depending on their signs. When the energy input to the layer exceeds the outgoing energy, there is a flux convergence ($\Delta H_S > 0$) which results in the warming of the layer. On the other hand, when energy going out exceeds that coming in, the layer cools as a result of flux divergence ($\Delta H_S < 0$). In special circumstances of energy coming

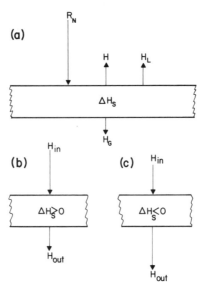

Fig. 2.2   Schematic representation of (a) energy budget of a layer, (b) flux convergence, and (c) flux divergence.

in exactly balancing the energy going out, there is no change in the energy storage of the layer ($\Delta H_S = 0$) or its temperature with time. These processes of vertical flux convergence and divergence are schematically shown in Fig. 2.2b and c.

2.2.3   ENERGY BUDGET OF A CONTROL VOLUME

In the energy budgets of an extensive horizontal surface and interfacial layer, only the vertical fluxes of energy are involved. When the surface under consideration is not flat and horizontal or when there are significant changes (advections) of energy fluxes in the horizontal, then it would be more appropriate to consider the energy budget of a control volume. In principle, this is similar to the energy budget of a layer, but now one must consider the various energy fluxes integrated or averaged over the bounding surface of the control volume. The energy budget equation for a control volume can be expressed as

$$\bar{R}_N = \bar{H} + \bar{H}_L + \bar{H}_G + \Delta H_S \qquad (2.4)$$

where the overbar over a flux quantity denotes its average value over the entire area (A) of the bounding surface and the rate of storage is given by

$$\Delta H_S = \frac{1}{A} \int \frac{\partial}{\partial t} (\rho c T)\, dV \tag{2.5}$$

in which the integration is over the control volume V.

In practice, detailed measurements of energy fluxes are rarely available in order to evaluate their net contributions to the energy budget of a large irregular volume. Still, with certain simplifying assumptions, such energy budgets have been used in the context of several large field experiments over the oceans, such as the 1969 Barbados Oceanographic and Meteorological Experiment (BOMEX) and the 1975 Air Mass Transformation Experiment (AMTEX).

The concept of flux convergence or divergence and its relation to warming or cooling of the medium is more general than depicted in Fig. 2.2 for a horizontal layer. For a control volume, the direction of flux is unimportant. According to Eq. (2.4), the net convergence or divergence of energy fluxes in all directions determines the rate of energy storage and, hence, the rate of warming or cooling of the medium in the control volume (see Oke, 1978, Chapter 2).

2.3 SOME EXAMPLES OF ENERGY BUDGET

A few cases of observed energy budgets will be discussed here for illustrative purposes only. It should be pointed out that relative magnitudes of the various terms in the energy budget may differ considerably for other places, times, and weather conditions.

2.3.1 BARE GROUND

Measured energy fluxes over a dry lake bed (desert) on a hot summer day are shown in Fig. 2.3. This represents the simplest case of energy balance [Eq. (2.1)] for a flat, dry, and bare surface in the absence of any evaporation or condensation ($H_L = 0$). It is also an example of thermally extreme climatic environment; the observed maximum difference in the temperatures between the surface and the air at 2 m was about 28°C.

Over a bare, dry ground the net radiation is entirely balanced by direct heat exchanges with the surface by both the air and soil media. However, the relative magnitudes of these fluxes may change considerably from day to night, as indicated by measurements in Fig. 2.3.

Wetting of the subsurface soil by precipitation or artificial irrigation can dramatically alter the surface energy balance, as well as the microclimate near the surface. The daytime net radiation for the same latitude, season, and weather conditions becomes larger for the wet surface, because of the

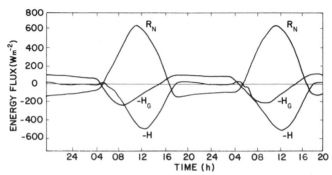

Fig. 2.3 Observed diurnal energy balance over a dry lake bed at El Mirage, California, on June 10 and 11, 1950. [After Vehrencamp (1953).]

reduced albedo (reflectivity) and increased absorption of shortwave radiation by the surface. The latent heat flux (H_L) becomes an important or even dominant component of the surface energy balance [Eq. (2.1)], while the sensible heat flux to air (H) is considerably reduced. The latter may even become negative during early periods of irrigation, particularly for small areas, due to advective effects. This is called the "oasis effect," which disappears as irrigation is applied over large areas and horizontal advections become less important. At later times, the relative importance of H and H_L will depend on the soil moisture content and soil temperature at the surface. As the soil dries up, evaporation rate (E) and $H_L = L_e E$ decrease, while the daytime surface temperature and the sensible heat flux increase with time, for the same environmental conditions. Thus, the daytime Bowen ratio is expected to increase with time after precipitation or irrigation.

2.3.2 VEGETATIVE SURFACES

The growth of vegetation over an otherwise flat surface introduces several complications into the energy balance. First, the ground surface is no longer the most appropriate datum for the surface energy balance, because the radiative, sensible, and latent heat fluxes are all spatially variable within the vegetative canopy. The energy budget of the whole canopy layer [Eq. (2.2)] will be more appropriate to consider. For this, measurements of R_N, H, and H_L are needed at the top of the canopy (preferably, well above the tops of plants or trees where horizontal variations of fluxes may be neglected).

Second, the rate of energy storage (ΔH_S) consists of two parts, namely, the rate of physical heat storage and the rate of biochemical heat storage as a result of photosynthesis and carbon dioxide exchange. The latter may

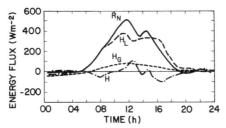

Fig. 2.4 Observed diurnal energy budget of a barley field at Rothamsted, England, on July 23, 1963. [From Oke (1987); after Long *et al.* (1964).]

not be important on time scales of few hours to a day, commonly used in micrometeorology. Nevertheless, the rate of heat storage by a vegetative canopy is not easy to measure or calculate.

Third, the latent heat exchange occurs not only due to evaporation or condensation at the surface, but to a large extent due to transpiration from the plant leaves. The combination of evaporation and transpiration is called evapotranspiration; it produces nearly a constant flux of water vapor above the canopy layer.

An example of the observed energy budget over a barley field on a summer day in England is shown in Fig. 2.4. Note that the latent heat flux due to evapotranspiration is the dominant energy component, which approximately balances the net radiation, while H and H_G are an order of magnitude smaller. In the late afternoon and evening, H_L even exceeded the net radiation and H became negative (downward heat flux). The rate of heat storage was not measured, but is estimated (from energy balance) to be small in this case.

Forest canopies have similar features as plant canopies, aside from the obvious differences in their sizes and stand architectures. Much larger heights of trees and the associated biomass of the forest canopy suggest that the rate of heat storage may not be insignificant, even over short periods of the order of a day.

A typical example of the observed energy budget of a Douglas fir canopy is given in Fig. 2.5. In this case, ΔH_S was roughly determined from the estimates of biomass and heat capacity of the trees and measurements of air temperatures within the canopy and, then, added to the measured ground heat flux to get the combined $H_G + \Delta H_S$ term. This term is relatively small during the daytime, but of the same order of magnitude as the net radiation (R_N) at night. Note that, during daytime, R_N is more or less equally partitioned between the sensible and latent heat fluxes to the air. For other examples of measured energy budgets over forests, the reader may refer to Chapter 17 of Munn (1966).

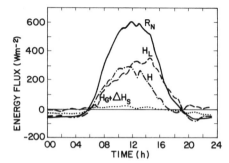

Fig. 2.5   Observed energy budget of a Douglas fir canopy at Haney, British Columbia, on July 23, 1970. [From Oke (1987); after McNaughton and Black (1973).]

2.3.3 WATER SURFACES

Water covers more than two-thirds of the earth's surface. Therefore, it is important to understand the energy budget of water surfaces. It is complicated, however, by the fact that water is a fluid with a dynamically active surface and a surface boundary layer or mixed layer in which the motions are generally turbulent. Thus, convective and advective heat transfers within the surface boundary layer in water essentially determine H_G. However, it is not easy to measure these transfers or H_G directly. As discussed earlier, the radiation balance at the water surface is also complicated by the fact that shortwave radiation penetrates to considerable depth in water. Therefore, it will be more appropriate to consider the energy budget of a layer of water extending to a depth where both the convective and radiative heat exchanges become negligible. This will not be feasible, however, for shallow bodies of water with depths less than about 10 m.

Even over large lakes and oceans, simultaneous measurements of all the energy flux terms in Eq. (2.2) on a short-term basis are lacking, largely because of the experimental difficulties associated with floating platforms and sea spray. The sensible and latent heat fluxes are the most frequently measured or estimated quantities. Over most ocean areas, the latter dominates the former and the Bowen ratio is usually much less than unity (it may approach unity during periods of intense cold air advection over warm waters). The rate of heat storage in the oceanic mixed layer plays an important role in the energy budget. This layer acts as a heat sink ($\Delta H_S > 0$) by day and a heat source ($\Delta H_S < 0$) at night.

The significance of net radiation in the energy budget over water is not so clear. For some measurements of heat balance at sea on a daily basis in the course of a large experiment, the reader may refer to Kondo (1976).

2.4 APPLICATIONS

The following list includes some of the applications of energy balance at or near the earth's surface:

- Prediction of surface temperature and frost conditions
- Indirect determination of the surface fluxes of heat (sensible and latent) to or from the atmosphere
- Estimation of the rate of evaporation from bare ground and water surfaces and evapotranspiration from vegetative surfaces
- Estimation of the rate of heat storage or loss by an oceanic mixed layer or a vegetative canopy
- Study of microclimates of the various surfaces
- Prediction of icing conditions on highways

In all these and possibly other applications, the various terms in the appropriate energy balance equation, except for the one to be estimated or predicted, have to be measured, calculated, or parameterized in terms of other measured quantities.

PROBLEMS AND EXERCISES

1. Describe the typical conditions in which you would expect the Bowen ratio to be as follows:
   (a) Much less than unity
   (b) Much greater than unity
   (c) Negative
2. (a) Over the tropical oceans the Bowen ratio is typically 0.1. Estimate the sensible and latent heat fluxes to the atmosphere, as well as the rate of evaporation, in millimeters per day, from the ocean surface, when the net radiation received just above the surface is 400 W m^{-2}, the heat flux to the water below 50 m is negligible, the rate of warming of the 50-m-deep oceanic mixed layer is 0.05°C day^{-1}, and the sea surface temperature is 30°C.
   (b) What will be the rate of warming or cooling of the 50-m-deep oceanic mixed layer in a region of intense cold-air advection where the Bowen ratio is estimated to be 0.5, the net radiation loss from the surface is 50 W m^{-2}, and the rate of evaporation is 20 mm day^{-1}?
3. Giving schematic depictions, briefly discuss the energy balance of an extensive, uniform snowpack for the following:

   (a) Below-freezing air temperatures
   (b) Above-freezing air temperatures
4. (a) Give a derivation of Eq. (2.3) for the rate of heat storage in a soil layer.
   (b) Will the same expression apply to an oceanic mixed layer? Give reasons for your answer.
5. Explain the following terms or concepts used in connection with the energy balance near the surface:
   (a) "Ideal" surface
   (b) Evaporative cooling
   (c) Oasis effect
   (d) Flux divergence
6. What are the major differences between the energy budgets of a bare soil surface and a vegetative surface?

Chapter 3 | Radiation Balance near the Surface

3.1 RADIATION LAWS AND DEFINITIONS

The transfer of energy by rapid oscillations of electromagnetic fields is called radiative transfer or simply radiation. These oscillations may be considered as traveling waves characterized by their wavelength λ or wave frequency c_m/λ, where c_m is the wave speed in a given medium. All electromagnetic waves travel at the speed of light $c \cong 3 \times 10^8$ m sec^{-1} in empty space and nearly the same speed in air ($c_m \cong c$). There is an enormous range or spectrum of electromagnetic wavelengths or frequencies. Here we are primarily interested in the approximate range 0.1–100 μm, in which significant contributions to the radiation balance of the atmosphere or the earth's surface occur. This represents only a tiny part of the entire electromagnetic wave spectrum. Of this, the visible light constitutes a very narrow range of wavelengths (0.40–0.76 μm).

The radiant flux density, or simply the radiative flux, is defined as the amount of radiant energy (integrated over all wavelengths) received at or emitted by a unit area of the surface per unit time. The SI unit of radiative flux is W m^{-2}, which is related to the CGS unit cal cm^{-2} min^{-1}, commonly used earlier in meteorology as 1 cal cm^{-2} min$^{-1} \cong 698$ W m^{-2}.

3.1.1 BLACKBODY RADIATION LAWS

Any body having a temperature above absolute zero emits radiation. If a body at a given temperature emits the maximum possible radiation per unit area of its surface, per unit time, at all wavelengths, it is called a perfect radiator or "blackbody." The flux of radiation (R) emitted by such a body is given by the *Stefan–Boltzmann law*

$$R = \sigma T^4 \tag{3.1}$$

where σ is the Stefan–Boltzmann constant = 5.67×10^{-8} W m^{-2} K^{-4}, and T is the surface temperature of the body in absolute (K) units.

Planck's law expresses the radiant energy per unit wavelength emitted by a blackbody as a function of its surface temperature

$$R_\lambda = (2\pi h_p c^2/\lambda^5) \, [\exp(h_p c/b\lambda T) - 1]^{-1} \tag{3.2}$$

where h_p is Planck's constant $= 6.626 \times 10^{-34}$ J sec, and b is the Boltzmann constant $= 1.381 \times 10^{-23}$ J K^{-1}. Note that the total radiative flux is given by

$$R = \int_0^\infty R_\lambda \, d\lambda \tag{3.3}$$

Equation (3.2) can be used to calculate and compare the spectra of blackbody radiation at various body surface temperatures. The wavelength at which R_λ is maximum turns out to be inversely proportional to the absolute temperature and is given by *Wien's law*

$$\lambda_{max} = 2897/T \tag{3.4}$$

when λ_{max} is expressed in micrometers.

From the above laws, it is clear that the radiative flux emitted by a blackbody varies in proportion to the fourth power of its surface temperature, while the wavelengths making the most contribution to the flux, especially λ_{max}, change inversely proportional to T.

3.1.2   SHORTWAVE AND LONGWAVE RADIATIONS

The spectrum of solar radiation received at the top of the atmosphere is well approximated by the spectrum of a blackbody having a surface temperature of about 6000 K. Thus, sun may be considered as a blackbody with an equivalent surface temperature of about 6000 K and $\lambda_{max} \cong 0.48 \, \mu$m.

The observed spectra of terrestrial radiation near the surface, particularly in the absence of absorbing substances such as water vapor and carbon dioxide, can also be approximated by the equivalent blackbody radiation spectra given by Eq. (3.2). However, the equivalent blackbody temperature (T_{eb}) is expected to be smaller than the actual surface temperature (T); the difference, $T - T_{eb}$, depends on the radiative characteristics of the surface, which will be discussed later.

Idealized or equivalent blackbody spectra of solar ($T_{eb} = 6000$ K) and terrestrial ($T_{eb} = 287$ K) radiation, both normalized with respect to their peak flux per unit wavelength, are compared in Fig. 3.1. Note that almost all the solar energy flux is confined to the wavelength range of 0.15–4.0 μm, while the terrestrial radiation is mostly confined to the range of 3–100 μm. Thus, there is very little overlap between the two spectra, with their

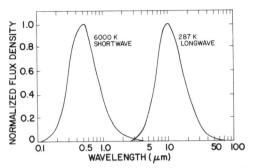

Fig. 3.1 Calculated blackbody spectra of (flux density of) solar and terrestrial radiation, both normalized by their peak flux density. [From Rosenberg *et al.* (1983).]

peak wavelengths (λ_{max}) separated by nearly a factor of 20. In meteorology, the above two ranges of wavelengths are characterized as shortwave and longwave radiations, respectively.

3.1.3 RADIATIVE PROPERTIES OF NATURAL SURFACES

Natural surfaces are not perfect radiators or blackbodies, but are, in general, gray bodies. They are generally characterized by several different radiative properties, which are defined as follows.

Emissivity is defined as the ratio of the energy flux emitted by the surface at a given wavelength and temperature to that emitted by a blackbody at the same wavelength and temperature. In general, the emissivity may depend on the wavelength and will be denoted by ε_λ. For a blackbody, $\varepsilon_\lambda = 1$ for all wavelengths.

Absorptivity is defined as the ratio of the amount of radiant energy absorbed by the surface material to the total amount of energy incident on the surface. In general, absorptivity is also dependent on wavelength and will be denoted by α_λ. A perfect radiator is also a perfect absorber of radiation, so that $\alpha_\lambda = 1$, for a blackbody.

Reflectivity is defined as the ratio of the amount of radiation reflected to the total amount incident upon the surface and will be denoted by r_λ.

Transmissivity is defined as the ratio of the radiation transmitted to the subsurface medium to the total amount incident upon the surface and will be denoted by t_λ.

It is clear from the above definitions that

$$\alpha_\lambda + r_\lambda + t_\lambda = 1 \tag{3.5}$$

so that absorptivity, reflectivity, and transmissivity must have values between zero and unity.

Kirchoff's law states that for a given wavelength, absorptivity of a material is equal to its emissivity, i.e.,

$$\alpha_\lambda = \varepsilon_\lambda \tag{3.6}$$

Natural surfaces are in general radiatively "gray," but in certain wavelength bands their emissivity or absorptivity may be close to unity. For example, in the so-called atmospheric window (8- to 14-μm wavelength band), water, wet soil, and vegetation have emissivities of 0.97–0.99.

In studying the radiation balance for the energy budget near the earth's surface, we are more interested in the radiative fluxes integrated over all the wavelengths of interest, rather than in their wavelength-dependent spectral decomposition. For this an overall or integrated emissivity and an integrated reflectivity, pertaining to radiation in certain range of wavelengths, are used to characterize the surface. In particular, the term "albedo" is used to represent the integrated reflectivity of the surface for shortwave (0.15–4 μm) radiation, while the overall emissivity (ε) of the surface refers primarily to the longwave (3–100 μm) radiation. For natural surfaces, the emission of shortwave radiation is usually neglected, and the emission of longwave radiative flux is given by the modified Stefan–Boltzmann law

$$R_L = -\varepsilon\sigma T^4 \tag{3.7}$$

where the negative sign is introduced in accordance with our sign convention for radiative fluxes. Typical values of surface albedo (a) and emissivity (ε) for different types of surfaces are given in Table 3.1.

The albedo of water is quite sensitive to the sun's zenith angle and may approach unity when the sun is near the horizon (rising or setting sun). It is also dependent on the wave height or sea state. To a lesser extent, the solar zenith angle is also found to affect albedos of soil, snow, and ice surfaces. Although snow has a high albedo, deposition of dust and aerosols, including man-made pollutants such as soot, can significantly lower the albedo of snow. The albedo of a soil surface is strongly dependent on the wetness of the soil.

Infrared emissivities of natural surfaces do not vary over a wide range; most surfaces have emissivities greater than 0.9, while only in a few cases values become less than 0.8. Green, lush vegetation and forests are characterized by highest emissivities, approaching close to unity.

3.2   SHORTWAVE RADIATION

The ultimate source of all shortwave radiation received at or near the earth's surface is the sun. A large part of it comes directly from the sun,

Table 3.1

Radiative Properties of Natural Surfaces[a]

Surface type	Other specifications	Albedo (a)	Emissivity (ε)
Water	Small zenith angle	0.03–0.10	0.92–0.97
	Large zenith angle	0.10–0.50	0.92–0.97
Snow	Old	0.40–0.70	0.82–0.89
	Fresh	0.45–0.95	0.90–0.99
Ice	Sea	0.30–0.40	0.92–0.97
	Glacier	0.20–0.40	
Bare sand	Dry	0.35–0.45	0.84–0.90
	Wet	0.20–0.30	0.91–0.95
Bare soil	Dry clay	0.20–0.35	0.95
	Moist clay	0.10–0.20	0.97
	Wet fallow field	0.05–0.07	
Paved	Concrete	0.17–0.27	0.71–0.88
	Black gravel road	0.05–0.10	0.88–0.95
Grass	Long (1 m) Short (0.02 m)	0.16–0.26	0.90–0.95
Agricultural	Wheat, rice, etc.	0.10–0.25	0.90–0.99
	Orchards	0.15–0.20	0.90–0.95
Forests	Deciduous	0.10–0.20	0.97–0.98
	Coniferous	0.05–0.15	0.97–0.99

[a] Compiled from Sellers (1965), Kondratyev (1969), and Oke (1978).

while other parts come in the forms of reflected radiation from the surface and clouds and scattered radiation from atmospheric particulates or aerosols.

3.2.1 SOLAR RADIATION

As a measure of the intensity of solar radiation, the solar constant (S_0) is defined as the flux of solar radiation falling on a surface normal to the solar beam at the outer edge of the atmosphere, when the earth is at its mean distance from the sun. An accurate determination of its value has been sought by many investigators. The best estimate of $S_0 = 1368$ W m^{-2} is based on a series of recent measurements from high-altitude platforms such as the Solar Maximum Satellite; other values proposed in the literature fall in the range 1350–1400 W m^{-2}. There are speculations that the solar constant may vary some with changes in the sun spot activity, which has a predominant cycle of about 11 years. There are also suggestions of very long-term variability of the solar constant which may have been partially responsible for long-term climatic fluctuations in the past.

The actual amount of solar radiation received at a horizontal surface per unit area over a specified time is called insolation. It depends strongly on the solar zenith angle γ and also on the ratio (d/d_m) of the actual distance to the mean distance of the earth from the sun. The combination of the so-called inverse-square law and Lambert's cosine law gives the flux density of solar radiation at the top of the atmosphere as

$$R_0 = S_0 \, (d_m/d)^2 \cos \gamma \qquad (3.8)$$

Then, insolation for a specified period of time between t_1 and t_2 is given by

$$I_0 = \int_{t_1}^{t_2} R_0(t) \, dt \qquad (3.9)$$

Thus, one can determine the daily insolation from Eq. (3.9) by integrating the solar flux density with time over the daylight hours.

For a given calendar day/time and latitude, the solar zenith angle (γ) and the ratio d_m/d can be determined from standard astronomical formulas or tables, and the solar flux density and insolation at the top of the atmosphere can be evaluated from Eqs. (3.8) and (3.9). These are also given in Smithsonian Meteorological Tables.

The solar flux density (R_s) and insolation (I) received at the surface of the earth may be considerably smaller than their values at the top of the atmosphere, because of the depletion of solar radiation in passing through the atmosphere. The largest effect is that of clouds, especially low stratus clouds. In the presence of scattered moving clouds, R_s becomes highly variable.

The second important factor responsible for the depletion of solar radiation is atmospheric turbidity, which refers to any condition of the atmosphere, excluding clouds, which reduces its transparency to shortwave radiation. The reduced transparency is primarily due to the presence of particulates, such as pollen, dust, smoke, and haze. Turbidity of the atmosphere in a given area may result from a combination of natural sources, such as wind erosion, forest fires, volcanic eruptions, sea spray, etc., and various man-made sources of aerosols. Particles in the path of a solar beam reflect a part of radiation and scatter the other part. Large, solid particles reflect more than they scatter light, affecting all visible wavelengths equally. Therefore, the sky appears white in the presence of these particles. Even in an apparently clear atmosphere, air molecules and very small (submicrometer size) particles scatter the sun's rays. According to Rayleigh's scattering law, scattering varies inversely as the fourth power of the wavelength. Consequently, the sky appears blue because there is preferred scattering of blue light (lowest wavelengths of the visible spectrum) over other colors.

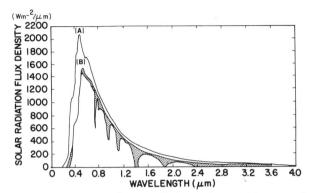

Fig. 3.2 Observed flux density of solar radiation at the top of the atmosphere (curve A) and at sea level (curve B). The shaded areas represent absorption due to various gases in a clear atmosphere. [From Liou (1983).]

Even in a cloud-free nonturbid atmosphere, atmospheric gases, such as oxygen, ozone, carbon dioxide, water vapor, and nitrous oxides, absorb parts of solar radiation in selected wavelength bands. For example, much of the ultraviolet radiation is absorbed by stratospheric ozone and oxygen, and water vapor and carbon dioxide have a number of absorption bands at wavelengths larger than 0.8 μm.

The combined effects of scattering and absorption of solar radiation in a clear atmosphere can be seen in Fig. 3.2. Here, curve A represents the measured spectrum of solar radiation at the top of the atmosphere, curve B represents the measured spectrum at the sea level, and the shaded areas represent the absorption of energy by various gases, primarily O_3, O_2, CO_2, and H_2O. The unshaded area between curve A and the outer envelope of the shaded area represents the depletion of solar radiation due to scattering.

3.2.2 REFLECTED RADIATION

A significant fraction (depending on the surface albedo) of the incoming shortwave radiation is reflected back by the surface. Shortwave reflectivities or albedos for the various natural and man-made surfaces are given in Table 3.1. Knowledge of surface albedos and of their possible changes (seasonal, as well as long term) due to man's activities, such as deforestation, agriculture, and urbanization, has been of considerable interest to climatologists. Satellites now permit systematic and frequent observations of surface albedos over large regions of the world. Any inadvertent or intentional changes in the local, regional, or global albedo may cause significant changes in the surface energy balance and hence in the micro- or macroclimate. For example, deliberate modification of albedo by large-scale surface covering has been proposed as a means of increasing precipitation in certain arid regions. It has also been suggested that large, frequent oil spills in the ice- and snow-covered arctic region could lead to significant changes in the regional energy balance and climate.

Snow is the most effective reflector (high albedo) of shortwave radiation. On the other hand, water is probably the poorest reflector (low albedo), while ice falls between snow and water. Since large areas of earth are covered by water, sea ice, and snow, significant changes in the snow and ice cover are likely to cause perceptible changes in the regional and, possibly, global albedo.

Most bare rock, sand, and soil surfaces reflect 10–45% of the incident shortwave radiation, the highest value being for the desert sands. Albedos of most vegetative surfaces fall in the range of 10–25%. Wetting of the surface by irrigation or precipitation considerably lowers the albedo, while snowfall increases it. Albedos of vegetative surfaces, being sensitive to the solar elevation angle, also show some diurnal variations, with their minimum values around noon and maximum values near sunrise and sunset (see, e.g., Rosenberg *et al.*, 1983, Chapter 1).

Shortwave reflectivities of clouds vary over a wide range, depending on the cloud type, height, and size, as well as on the angle of incidence of radiation (see, e.g., Welch *et al.*, 1980). Thick stratus, stratocumulus, and nimbostratus clouds are good reflectors (a = 0.6–0.8, at normal incidence), large cumulus clouds are moderate reflectors (a = 0.2–0.5), and small cumuli are poor reflectors (a < 0.2). Only a small part of solar radiation may reach the ground in cloudy conditions. A significant part of the reflected radiation from the surface may be reflected back by clouds. In fact, shortwave radiation is likely to undergo multiple reflections between the surfaces and bases of clouds, thereby increasing the effective albedo of the surface.

3.2.3 DIFFUSE RADIATION

Diffuse or sky radiation is that portion of the solar radiation that reaches the earth's surface after having been scattered by molecules and suspended particulates in the atmosphere. In cloudy conditions it also includes the portion of the shortwave radiation which is reflected by the clouds. It is the incoming shortwave radiation in shade. Before sunrise and after sunset, all shortwave radiation is in diffuse form. The ratio of diffuse to the total incoming shortwave radiation varies diurnally, seasonally, and with latitude. In high latitudes, diffuse radiation is very important; even in midlatitudes it constitutes 30–40% of the total incoming solar radiation. Cloudiness considerably increases the ratio of diffuse to total solar radiation.

3.3   LONGWAVE RADIATION

The longwave radiation R_L received at or near the surface has two components: (1) outgoing radiation $R_{L\uparrow}$ from the surface, and (2) incoming radiation $R_{L\downarrow}$ from the atmosphere, including clouds. These are discussed separately in the following sections.

3.3.1   TERRESTRIAL RADIATION

All natural surfaces radiate energy, depending on their emissivities and surface temperatures, whose flux density is given by Eq. (3.7). As discussed earlier in Section 3.1, emissivities for most natural surfaces range from 0.9 to 1.0. Thus knowing the surface temperature and a crude estimate of emissivity, one can determine terrestrial radiation from Eq. (3.7) to better than 10% accuracy. Difficulties arise, however, in measuring or even defining the surface temperature, especially for vegetative surfaces. In such cases, it may be more appropriate to determine the apparent surface temperature, or equivalent blackbody temperature T_{eb} from measurements of terrestrial radiation near the surface using Eq. (3.1).

In passing through the atmosphere, a large part of the terrestrial radiation is absorbed by atmospheric gases, such as water vapor, carbon dioxide, nitrogen oxides, methane, and ozone. In particular, water vapor and CO_2 are primarily responsible for absorbing the terrestrial radiation and reducing its escape to the space (the so-called greenhouse effect).

3.3.2   ATMOSPHERIC RADIATION

From our discussion earlier, it is clear that the atmosphere absorbs much of the longwave terrestrial radiation and a significant part of the solar radiation. The atmospheric gases and aerosols which absorb energy also radiate energy, depending on the vertical distributions of their concentrations or mixing ratios and air temperature as functions of height. An important aspect of the atmospheric radiation is that absorption and emission of radiation by various gases occur in a series of discrete wavelengths or bands of wavelengths, rather than continuously across the spectrum, as shown in Fig. 3.3. All layers of the atmosphere participate, to varying degrees, in absorption and emission of radiation, but the atmospheric

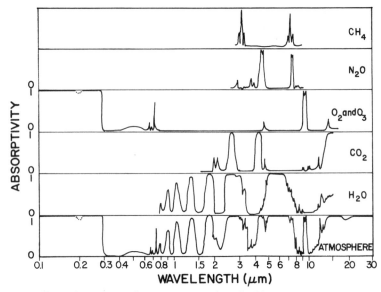

Fig. 3.3 Absorption spectra of water vapor, carbon dioxide, oxygen and ozone, nitrogen oxide, methane, and the atmosphere. [From Fleagle and Businger (1980).]

boundary layer is most important in this, because the largest concentrations of water vapor, CO_2, and other gases occur in this layer.

Clouds, when present, are the major contributors to the incoming longwave radiation to the surface. They radiate like blackbodies ($\varepsilon \cong 1$) at their respective cloud base temperatures. However, some of the radiation is absorbed by water vapor, CO_2, and other absorbing gases before reaching the earth's surface.

Computation of incoming longwave radiation from the atmosphere is tedious and complicated, even when the distributions of water vapor, CO_2, cloudiness, and temperature are measured. It is preferable to measure $R_{L\downarrow}$ directly with an appropriate radiometer.

3.4 RADIATION BALANCE NEAR THE SURFACE

The net radiation flux R_N in Eqs. (2.1) and (2.2) is a result of radiation balance between shortwave (R_S) and longwave (R_L) radiations at or near the surface, which can be written as

$$R_N = R_S + R_L \qquad (3.10)$$

Further, expressing shortwave and longwave radiation balance terms as

$$R_S = R_{S\downarrow} + R_{S\uparrow} \qquad (3.11)$$

$$R_L = R_{L\downarrow} + R_{L\uparrow} \qquad (3.12)$$

the overall radiation balance can also be written as

$$R_N = R_{S\downarrow} + R_{S\uparrow} + R_{L\downarrow} + R_{L\uparrow} \qquad (3.13)$$

where the downward and upward arrows denote incoming and outgoing radiation components, respectively.

The incoming shortwave radiation ($R_{S\downarrow}$) consists of both the direct-beam solar radiation and the diffuse radiation. It is also called insolation at the ground and can be easily measured by a solarimeter. It has strong diurnal variation (almost sinusoidal) in the absence of fog and clouds.

The outgoing shortwave radiation ($R_{S\uparrow}$) is actually the fraction of $R_{S\uparrow}$ that is reflected by the surface, i.e.,

$$R_{S\uparrow} = -aR_{S\downarrow} \qquad (3.14)$$

where a is the surface albedo. Thus, for a given surface, the net short-wave radiation $R_S = (1 - a)R_{S\downarrow}$ is essentially determined by insolation at the ground.

The incoming longwave radiation ($R_{L\downarrow}$) from the atmosphere, in the absence of clouds, depends primarily on the distributions of temperature, water vapor, and carbon dioxide. It does not show a significant diurnal variation. The outgoing terrestrial radiation ($R_{L\uparrow}$), being proportional to the fourth power of the surface temperature in absolute units, shows stronger diurnal variation, with its maximum value in the early afternoon and minimum value at dawn. The two components are usually of the same order of magnitude, so that the net longwave radiation (R_L) is generally a small quantity.

Under clear skies, $|R_L| \ll R_S$ during the bright daylight hours and an approximate radiation balance

$$R_N \cong R_S = (1 - a)R_{S\downarrow} \qquad (3.15)$$

which can be used to determine net radiation from simpler measurements or calculation of solar radiation at the surface. At nighttime, however, $R_{S\downarrow} = 0$, and the radiation balance becomes

$$R_N = R_L = R_{L\downarrow} + R_{L\uparrow} \qquad (3.16)$$

Frequently at night $R_{L\downarrow} < -R_{L\uparrow}$, so that R_N or R_L is usually negative, implying radiative cooling of the surface.

Near the sunrise and sunset times, all the components of the radiation balance are of the same order of magnitude and Eq. (3.10) or (3.13) will be more appropriate than the simplified Eq. (3.15) or (3.16).

SOME EXAMPLES OF RADIATION BALANCE

For the purpose of illustration, some examples of measured radiation balance over different types of surfaces are provided. Figure 3.4 represents the various components of Eq. (3.10) for a clear August day in England over a thick stand of grass, when the diurnal variations of grass temperature were about 20°C. Note that in this case about one-fourth of the incident solar radiation was reflected back by the surface, while about three-fourths was absorbed. The net radiation is slightly less than R_S, even during the midday hours, because of the net loss due to longwave radiation. The diurnal variation of $R_{L\downarrow}$ is much less than that of $R_{L\uparrow}$ (28% variation in $R_{L\uparrow}$ is consistent with the observed diurnal range of 20°C in surface temperature).

Figure 3.5 shows the diurnal variation of radiation budget components over Lake Ontario on a clear day in August. Note that the measured shortwave radiation ($R_{S\downarrow}$) near the lake surface is only about two-thirds of the computed solar radiation (R_0) at the top of the atmosphere. Of this, 25–30% is in the form of diffuse-beam radiation (R_D) at midday and the

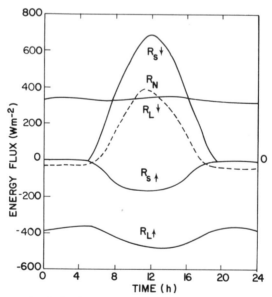

Fig. 3.4 Observed radiation budget over a 0.2-m-tall stand of native grass at Matador, Saskatchewan, on July 30, 1971. [From Oke (1987); after Ripley and Redmann (1976).]

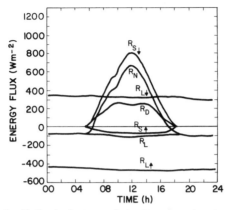

Fig. 3.5 Observed radiation budget over Lake Ontario under clear skies on August 28, 1969. [From Oke (1987); after Davies *et al.* (1970).]

rest as direct-beam solar radiation (not plotted). The outgoing shortwave radiation ($R_{S\uparrow}$) is relatively small, due to the low albedo ($a \cong 0.07$) of water. Both the incoming and outgoing longwave radiation components are relatively constant with time, due to small diurnal variations in the temperatures of the lake surface and the air above it. The net longwave radiation is a constant energy loss throughout the period of observations. The net radiation (R_N) is dominated by $R_{S\downarrow}$ during the day and is equal to R_L at night.

3.5 RADIATIVE FLUX DIVERGENCE

The concept of energy flux convergence or divergence and its relation to cooling or warming of a layer of the atmosphere or submedium has been explained in Chapter 1. Here, we discuss the significance of net radiative flux convergence or divergence to warming or cooling in the lowest layer of the atmosphere, namely, the PBL.

The rate of warming or cooling of a layer of air due to change of net radiation with height can be calculated from the principle of conservation of energy. Considering a thin layer between the levels z and $z + \Delta z$, where the net radiative fluxes are $R_N(z)$ and $R_N(z + \Delta z)$, as shown in Fig. 3.6, we get

$$\rho c_p \Delta z (\partial T/\partial t)_R = R_N(z + \Delta z) - R_N(z) = (\partial R_N/\partial z)\Delta z$$

or

$$(\partial T/\partial t)_R = (1/\rho c_p)(\partial R_N/\partial z) \qquad (3.17)$$

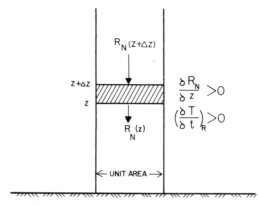

Fig. 3.6 Schematic of radiative flux convergence or divergence in the lower atmosphere.

where $(\partial T/\partial t)_R$ is the rate of change of temperature due to radiation and $\partial R_N/\partial z$ represents the convergence or divergence of net radiation. Radiative flux convergence occurs when R_N increases with height ($\partial R_N/\partial z > 0$) and divergence occurs when $\partial R_N/\partial z < 0$. The former leads to warming and the latter to cooling of the air.

In the daytime, during clear skies, net radiation is dominated by the net shortwave radiation (R_S), which usually does not vary with height in the PBL. Therefore, changes in air temperature with time due to radiation are insignificant or negligible; the observed daytime warming of the PBL is largely due to the convergence of sensible heat flux. This may not be true, however, in the presence of fog and clouds in the PBL, when radiative flux convergence or divergence may also become important.

At night, the net radiation is entirely due to the net longwave radiation (R_L). Both the terrestrial and atmospheric radiations and, hence, R_L generally vary with height, because the concentrations of water vapor, CO_2, and other gases that absorb and emit longwave radiation vary strongly with height in the PBL. Frequently, there is significant radiative flux divergence (which implies that radiative heat loss increases with height) within the PBL, especially during clear and calm nights. A significant part of cooling in the PBL may be due to radiation, while the remainder is due to the divergence of sensible heat flux. In many theoretical studies of the nocturnal boundary layer (NBL), radiative flux divergence has erroneously been ignored; its importance in the determination of thermodynamic structure of the NBL has recently been pointed out (Garratt and Brost, 1981).

In the presence of strong radiative flux divergence or convergence, radiation measurements at some height above the surface may not be representative of their surface values. Corrections for the flux divergence need to be applied to such measurements.

3.6 APPLICATIONS

The radiation balance at or near the earth's surface has the following applications:

- Determining climate near the ground
- Determining radiative properties, such as albedo and emissivity of the surface
- Determining net radiation, which is an important component of the surface energy budget
- Defining and determining the apparent (equivalent blackbody) surface temperature for a complex surface
- Parameterizing the surface heat fluxes to soil and air in terms of net radiation
- Determining radiative cooling or warming of the PBL

These and other applications are particularly important to the fields of agricultural meteorology and climatology.

PROBLEMS AND EXERCISES

1. The following measurements were made over a dry, bare field during a very calm spring night:

Outgoing longwave radiation from the surface $= 400$ W m^{-2}

Incoming longwave radiation from the atmosphere $= 350$ W m^{-2}

   (a) Calculate the apparent (equivalent blackbody) temperature of the surface.
   (b) Calculate the actual surface temperature if surface emissivity is 0.95.
   (c) Estimate the heat flux into the submedium, making appropriate assumptions about other fluxes.
2. (a) Estimate the combined sensible and latent heat fluxes from the surface to the atmosphere, given the following observations:

Incoming shortwave radiation $= 800$ W m^{-2}

Heat flux to the submedium $= 150$ W m^{-2}

Albedo of the surface $= 0.35$

(b) What would be the result if the surface albedo were to drop to 0.07 after irrigation?

3. Show that the variation of about 28% in terrestrial radiation in Fig. 3.4 is consistent with the observed range of 10–30°C in surface temperatures.

4. (a) Discuss the importance and consequences of the radiative flux divergence at night above a grass surface.

   (b) If the net longwave radiation fluxes at 1 and 10 m above the surface are -135 and -150 W m^{-2}, respectively, calculate the rate of cooling or warming in °C/hour due to radiation alone.

5. Explain the nature and causes of depletion of the solar radiation in passing through the atmosphere.

6. Discuss the consequences of the absorption of longwave radiation by atmospheric gases and the so-called greenhouse effect.

7. Discuss the merits of the proposition that net radiation R_N can be deduced from measurements of solar radiation $R_{S\downarrow}$ during the daylight hours, using the empirical expression

$$R_N = AR_{S\downarrow} + B$$

where A and B are constants. On what factors are A and B expected to depend?

8. The following measurements were made at night from a meteorological tower:

$$\text{Net radiation at the surface} = -125 \text{ W m}^{-2}$$

$$\text{Net radiation at the 200-m level} = -250 \text{ W m}^{-2}$$

$$\text{Sensible heat flux at the surface} = -150 \text{ W m}^{-2}$$

$$\text{Planetary boundary layer height} = 150 \text{ m}$$

Calculate the average rate of cooling in the PBL due to the following:
(a) The radiative flux divergence
(b) The sensible heat flux divergence

Chapter 4 | Soil Temperatures and Heat Transfer

4.1 SURFACE TEMPERATURE

An "ideal" surface, defined in Chapter 2, may be considered to have a uniform surface temperature (T_s) which varies with time only in response to the time-dependent energy fluxes at the surface. An uneven or nonhomogeneous surface is likely to have spatially varying surface temperature. The surface temperature at a given location is essentially given by the surface energy balance, which in turn depends on the radiation balance, atmospheric exchange processes in the immediate vicinity of the surface, presence of vegetation or plant cover, and thermal properties of the subsurface medium. Here, we distinguish between the real surface (skin) temperature and the near-surface air temperature. The latter is measured at standard meteorological stations at the screen height of 1–2 m.

The direct *in situ* measurement of surface temperature is made very difficult by the extremely large temperature gradients that commonly occur near the surface in both the air (temperature gradients of 10–20 K mm^{-1} in air are not uncommon very close to the surface) and the soil media, by the finite dimensions of the temperature sensor, and by the difficulties of ventilating and shielding the sensor when it is placed at the surface. Therefore, the surface temperature is often determined by extrapolation of measured temperature profiles in soil and air, with the knowledge of their expected theoretical behaviors. Another, perhaps better, method of determining the surface temperature, when emissivity of the surface is known, is through remote sensors, such as a downward-looking radiometer which measures the flux of outgoing longwave radiation from the surface and, hence, T_s, using the modified Stefan–Boltzmann Eq. (3.7). If the surface emissivity is not known with sufficient accuracy, this method will give the apparent (equivalent blackbody) surface temperature. In this way, surface or apparent surface temperatures are being routinely monitored by weather satellites. Measurements are made in narrow regions of the spectrum in which water vapor and CO_2 are transparent. The technique has proved to be fairly reliable for measuring sea-surface temperatures, but not so reliable for land-surface temperatures.

The times of minimum and maximum in surface temperatures as well as the diurnal range are of considerable interest to micrometeorologists. On clear days, the maximum surface temperature is attained typically an hour or two after the time of maximum insolation, while the minimum temperature is reached in early morning hours. The maximum diurnal range is achieved for a relatively dry and bare surface, under relatively calm air and clear skies. For example, on bare soil in summer, midday surface temperatures of 50–60°C are common in arid regions, while early morning temperatures may be only 10–20°C. The bare-surface temperatures also depend on the texture of the soil. Fine-textured soils (e.g., clay) have greater heat capacities, as compared to coarse soils (e.g., sand).

The presence of moisture at the surface and in the subsurface soil greatly moderates the diurnal range of surface temperatures. This is due to the increased evaporation from the surface, and also due to increased heat capacity and thermal conductivity of the soil. Over a free water surface, on the average, about 90% of the net radiation is utilized for evaporation. Over a wet, bare soil also a substantial part of net radiation goes into evaporation in the beginning, but this fraction reduces as the soil surface dries up. Increased heat capacity of the soil further slows down the warming of the upper layer of the soil in response to radiative heating of the surface. The ground heat flux is also reduced by evaporation.

The presence of vegetation on the surface also reduces the diurnal range of surface temperatures. Part of the incoming solar radiation is intercepted by plant surfaces, reducing the amount reaching the surface. Therefore surface temperatures during the day are uniformly lower under vegetation than over a bare soil surface. At night the outgoing longwave radiation is also partly intercepted by vegetation, but the latter radiates energy back to the surface. This slightly slows down the radiative cooling of the surface. Vegetation also enhances the latent heat exchange due to evapotranspiration. It increases turbulence near the surface which provides more effective exchanges of sensible and latent heat between the surface and the overlying air. The combined effect of all these processes is to significantly reduce the diurnal range of temperatures of vegetative surfaces.

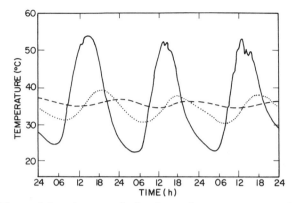

Fig. 4.1 Observed diurnal course of subsurface soil temperatures at various depths in a sandy loam with bare surface. ——, 2.5 cm; ·····, 15 cm; ---, 30 cm. [From Deacon (1969); after West (1952).]

4.2 SUBSURFACE TEMPERATURES

Subsurface soil temperatures are easier to measure than the surface temperature. It has been observed that the range or amplitude of diurnal variation of soil temperatures decreases exponentially with depth and becomes insignificant at a depth of the order of 1 m or less. An illustration of this is given in Fig. 4.1, which is based on measurements in a dry, sandy loam soil.

Soil temperatures depend on a number of factors, which also determine the surface temperature. The most important are the location (latitude) and the time of the year (month or season), net radiation at the surface, soil texture and moisture content, ground cover, and surface weather conditions. Temperature may increase, decrease, or vary nonmonotonically with depth, depending on the season and the time of the day.

In addition to the "diurnal waves" present in soil temperatures in the top layer, daily or weekly averaged temperatures show a nearly sinusoidal "annual wave" which penetrates to much greater depths (~10 m) in the soil. This is illustrated in Fig. 4.2.

A simple theory for explaining the observed diurnal and annual temperature waves in soils and variations of their amplitudes and phases with depth will be presented later.

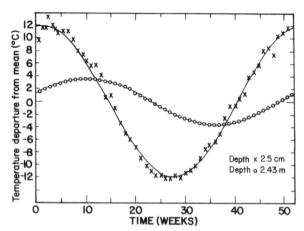

Fig. 4.2   Annual temperature waves in the weekly averaged subsurface soil temperatures at two depths in a sandy loam soil. ×, 2.5 cm; ○, 2.43 m. Fitted solid curves are sine waves. [From Deacon (1969); after West (1952).]

4.3 THERMAL PROPERTIES OF SOILS

Thermal properties relevant to the transfer of heat through a medium and its effect on the average temperature or distribution of temperatures in the medium are the mass density, specific heat, heat capacity, and thermal conductivity.

The specific heat (c) of a material is defined as the amount of heat absorbed or released in raising or lowering the temperature of a unit mass of the material by 1°. The product of mass density (ρ) and specific heat is called the heat capacity per unit volume, or simply heat capacity (C).

Through solid media and still fluids, heat is transferred primarily through conduction, which involves molecular exchanges. The rate of heat transfer or heat flux in a given direction is found to be proportional to the temperature gradient in that direction, i.e., the heat flux in the z direction

$$H = -k(\partial T/\partial z) \tag{4.1}$$

in which the proportionality factor k is known as the thermal conductivity of the medium. The ratio of thermal conductivity to heat capacity is called the thermal diffusivity α_h. Then, Eq. (4.1) can also be written as

$$H/\rho c = -\alpha_h(\partial T/\partial z) \tag{4.2}$$

Heat is transferred through soil, rock, and other subsurface materials (except for water in motion) by conduction, so that the above-mentioned molecular thermal properties also characterize the submedium. Table 4.1 gives typical values of these for certain soils, as well as for the reference air and water media, for comparison purposes.

Note that air has the lowest heat capacity, as well as the lowest thermal conductivity of all the natural materials, while water has the highest heat capacity. Thermal diffusivity of air is very large, because of its low density. Thermal properties of both air and water depend on temperature. Most soils consist of particles of different sizes and materials with a significant fraction of pore space, which may be filled with air and/or water. Thermal properties of soils, therefore, depend on the properties of the solid particles and their size distribution, porosity of the soil, and the soil moisture content. Of these, the soil moisture content is a short-term variable, changes of which may cause significant changes in heat capacity, conductivity, and diffusivity of soils (see, e.g., Oke, 1978, Chap. 2; Rosenberg et al., 1983, Chap. 2).

Addition of water to an initially dry soil increases its heat capacity and conductivity markedly, because it replaces air (a poor heat conductor) in pore space. Both the heat capacity and thermal conductivity are monotonic increasing functions of soil moisture content. Their ratio, thermal diffusivity, for most soils increases with moisture content initially, attains a maximum value, and then falls off with further increase in soil moisture content. It will be seen later that thermal diffusivity is a more appropriate measure of how rapidaly surface temperature changes are transmitted to deeper layers of submedium.

4.4 THEORY OF SOIL HEAT TRANSFER

Here we consider a uniform conducting medium (soil), with heat flowing only in the vertical direction. Let us consider the energy budget of an elemental volume consisting of a cylinder of horizontal cross-section area ΔA and depth Δz, bounded between the levels z and $z + \Delta z$, as shown in Fig. 4.3.

Heat flow in the volume of depth $z = H\Delta A$
Heat flow out of the volume at $z + \Delta z = [H + (\partial H/\partial z)\Delta z]\Delta A$
The net rate of heat flow in the control volume $= -(\partial H/\partial z)\Delta z\Delta A$
The rate of change of internal energy within the control volume
$\quad = (\partial/\partial t)(\rho\Delta A\Delta z cT)$

According to the law of conservation of energy, if there are no sources or sinks of energy within the elemental volume, the net rate of heat

Table 4.1

Molecular Thermal Properties of Natural Materials[a]

Material	Condition	Mass density ρ (kg m⁻³ × 10³)	Specific heat c (J kg⁻¹ K⁻¹ × 10³)	Heat capacity C (J m⁻³ K⁻¹ × 10⁶)	Thermal conductivity k (W m⁻¹ K⁻¹)	Thermal diffusivity α_h (m² sec⁻¹ × 10⁻⁶)
Air	20°C, Still	0.0012	1.00	0.0012	0.026	21.5
Water	20°C, Still	1.00	4.19	4.19	0.58	0.14
Ice	0°C, Pure	0.92	2.10	1.93	2.24	1.16
Snow	Fresh	0.10	2.09	0.21	0.08	0.38
Sandy soil	Dry	1.60	0.80	1.28	0.30	0.24
(40% pore space)	Saturated	2.00	1.48	2.98	2.20	0.74
Clay soil	Dry	1.60	0.89	1.42	0.25	0.18
(40% pore space)	Saturated	2.00	1.55	3.10	1.58	0.51
Peat soil	Dry	0.30	1.92	0.58	0.06	0.10
(80% pore space)	Saturated	1.10	3.65	4.02	0.50	0.12

[a] After Oke (1987).

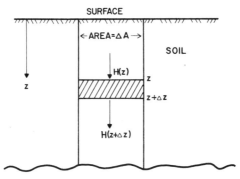

Fig. 4.3 Schematic of heat transfer in a vertical column of soil below a flat, horizontal surface.

flowing in volume should equal the rate of change of internal energy in the volume, so that

$$(\partial/\partial t)(\rho c T) = -\partial H/\partial z \qquad (4.2)$$

Further, assuming that the heat capacity of the medium does not vary with time, and substituting from Eq. (4.1) into (4.2), we obtain the Fourier's equation of heat conduction:

$$\partial T/\partial t = (\partial/\partial z)[(k/\rho c)(\partial T/\partial z)] = (\partial/\partial z)[\alpha_h(\partial T/\partial z)] \qquad (4.3)$$

The one-dimensional heat conduction equation derived here can easily be generalized to three dimensions by considering the net rate of heat flow in an elementary control volumn $\Delta x \Delta y \Delta z$ from all the directions. In our applications involving heat transfer through soils, however, we will be primarily concerned with the one-dimensional Eq. (4.2) or (4.3).

Equation (4.2) is often used to determine the ground heat flux H_G in the energy balance equation from measurements of soil temperatures as functions of time, at various depths below the surface. The method is based on the integration of Eq. (4.2) from $z = 0$ to D

$$H_G = H_D + \int_0^D \frac{\partial}{\partial t} (\rho c T) \, dz \qquad (4.4)$$

Where D is some reference depth where the soil heat flux H_D is either zero (e.g., if at $z = D$, $\partial T/\partial z = 0$) or can be easily estimated [e.g., using Eq. (4.1)]. The former is preferable whenever feasible, because it does not require the knowledge of thermal conductivity, which is more difficult to measure than heat capacity. Although, in principle, one can also use Eq. (4.1) to directly determine the ground heat flux, such a procedure would be highly unreliable in practice, because large errors are likely to be introduced in measurements of soil temperature gradient near the surface.

PROPAGATION OF THERMAL WAVE IN HOMOGENEOUS SOILS

The solution of Eq. (4.3), with given initial and boundary conditions, is used to study theoretically the propagation of thermal waves in a homogeneous soil or other submedium. For any arbitrary prescription of surface temperature as a function of time, the solution to Eq. (4.3) can be obtained numerically. Much about the physics of thermal wave propagation can be learned, however, from a simple analytic solution which is obtained when the surface temperature is specified as a sinusoidal function of time.

$$T_s = T_m + A_s \sin[(2\pi/P)(t - t_m)] \tag{4.5}$$

Where T_m is the mean temperature of the surface or submedium, A_s and P are the amplitude and period of the surface temperature wave, and t_m is the time when $T_s = T_m$, as the surface temperature is rising.

The solution of Eq. (4.3) satisfying the boundary conditions that at $z = 0$, $T = T_s(t)$, and as $z \to \infty$, $T \to T_m$, is given by

$$T = T_m + A_s \exp(-z/d) \sin[(2\pi/P)(t - t_m) - z/d] \tag{4.6}$$

which the reader may verify by substituting in Eq. (4.3). Here, d is the damping depth of the thermal wave, defined as

$$d = (P\alpha_h/\pi)^{1/2} \tag{4.7}$$

Note that the period of thermal wave in the soil remains unchanged, while its amplitude decreases exponentially with depth ($A = A_s \exp(-z/d)$); at $z = d$ the wave amplitude is reduced to about 37% of its value at the surface and at $z = 3d$ the amplitude decreases to about 5% of the surface value. The phase lag relative to the surface wave increases in proportion to depth (phase lag $= z/d$), so that there is a complete reversal of the wave phase at $z = \pi d$. The corresponding lag in the times of maximum or minimum in temperature is also proportional to depth (time lag $= zP/2\pi d$).

The results of the above simple theory would be applicable to the propagation of both the diurnal and annual temperature waves through a homogeneous submedium, provided that the thermal diffusivity of the medium remains constant over the whole period, and the surface temperature wave is nearly sinusoidal (the latter condition is usually not satisfied). Note that the damping depth, which is a measure of the extent of thermal wave propagation, for the annual wave is expected to be $\sqrt{365} = 19.1$ times the damping depth for the diurnal wave. For example, for a dry, sandy soil of thermal diffusivity $\alpha_h = 0.25 \times 10^{-6}\,\mathrm{m^2\,sec^{-1}}$, $d \simeq 0.082$ m for the diurnal wave, and $d \simeq 1.57$ m for the annual wave.

The observed temperature waves (see, e.g., Figs. 4.1 and 4.2) in bare, dry soils are found to conform well to the pattern predicted by the theory. In particular, the observed annual waves show nearly perfect sinusoidal forms. The plots of wave amplitude (on log scale) and phase or time lag as functions of depth (both on linear scale) are well represented by straight lines (see, e.g., Fig. 4.4) whose slopes determine the damping depth and, hence, thermal diffusivity of the soil. For example, from the amplitude data plotted in Fig. 4.4 one gets $\alpha_h \simeq 0.51 \times 10^{-6}\,\mathrm{m^2\,sec^{-1}}$; from the time-lag data, $\alpha_h \simeq 0.41 \times 10^{-6}\,\mathrm{m^2\,sec^{-1}}$; and from the amplitude decay of diurnal wave (Fig. 4.1) in the top 0.30 m, $\alpha_h \simeq 0.4 \times 10^{-6}\,\mathrm{m^2\,sec^{-1}}$, which are in good agreement (Deacon, 1969).

The ground heat flux can be obtained from Eq. (4.6) as

$$H_G = -k\left(\frac{\partial T}{\partial z}\right)_{z=0} = \left(2\pi\,\frac{\rho c k}{P}\right)^{1/2} A_s \sin\left[\frac{2\pi}{P}(t - t_m) + \frac{\pi}{4}\right] \qquad (4.8)$$

which predicts that the maximum in surface temperature should lag behind the maximum in ground heat flux by $P/8$, or 3 hr for the diurnal period and 1.5 months for the annual period. The above equation also indicates that the amplitude of ground heat flux wave is proportional to the square root of the product of heat capacity and thermal conductivity

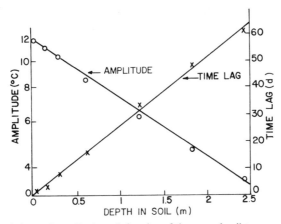

Fig. 4.4 Variations of amplitude and time lag of the annual soil temperature waves with depth in the soil. [From Deacon (1969).]

and inversely proportional to the square root of the period; it is also proportional to the amplitude of the surface temperature wave.

Actual conditions at the surface and within the submedium may differ considerably from the ideal conditions assumed in the simple theory of soil heat transfer presented here. For example, thermal properties may vary with depth because of the layered structure of the submedium, variations in the soil moisture content, and presence of roots and surface vegetation. Thermal properties may also vary with time in response to irrigation, precipitation, and evaporation. The diurnal variation of surface temperature often deviates from the sinusoidal form assumed in the simple theory. In such cases, numerical models of soil heat and moisture transfer with varying degrees of complication are used, in conjunction with the surface energy balance equation.

4.5 APPLICATIONS

Some of the applications of the knowledge of surface and soil temperatures and soil heat transfer are as follows:

- Study of surface energy budget and radiation balance
- Prediction of surface temperature and frost conditions
- Determination of the rate of heat storage or release by the submedium
- Study of microenvironment of plant cover, including the root zone
- Environmental design of underground structures
- Determination of the depth of permafrost zone in high latitudes

PROBLEMS AND EXERCISES

1. You want to measure the temperature of a bare soil surface using fine temperature sensors on a summer day when the sensible heat flux to air might be typically 600 W m^{-2} and the ground heat flux 100 W m^{-2}. Will you do this by extrapolation of measurements near the surface in air or in soil? Give reasons for your choice. Also estimate temperature gradients near the surface in both the air ($k = 0.03$ W m^{-1} K^{-1}) and the soil ($k = 0.30$ W m^{-1} K^{-1}).
2. Discuss the effect of vegetation or plant cover on the diurnal range of surface temperatures, as compared to that over a bare soil surface, and explain why green lawns are much cooler than roads and driveways on sunny afternoons.
3. What is the physical significance of the damping depth in the propaga-

tion of thermal waves in a submedium and on what factors does it depend?

4. What are the basic assumptions underlying the simple theory of soil heat transfer presented in this chapter? Discuss the situations in which these assumptions may not be valid.

5. By substituting from Eq. (4.6) into (4.3), verify that the former is a solution of Fourier's equation of heat conduction through a homogeneous soil medium, and then, derive the expression [Eq. (4.8)] for the ground heat flux.

6. Compare and contrast the alternative methods of determining the ground heat flux from measurements of soil temperatures and thermal properties.

7. The following soil temperatures (°C) were measured during the Great Plains Field Program at O'Neil, Nebraska:

Time (hr)	Day	Depth (m)				
		0.025	0.05	0.10	0.20	0.40
0435	August 31	25.54	26.00	26.42	26.32	24.50
0635		24.84	25.30	25.84	25.97	24.48
0835		25.77	25.35	25.42	25.56	24.36
1035		29.42	27.36	25.98	25.39	24.34
1235		33.25	30.32	27.62	25.57	24.27
1435		35.25	32.63	29.52	26.11	24.24
1635		34.84	33.20	30.62	26.88	24.26
1835		32.63	32.05	30.62	27.41	24.32
2035		30.07	30.20	29.91	27.68	24.47
2235		28.42	28.74	28.84	27.57	24.64
0035	September 1	27.09	27.50	27.84	27.22	24.73
0235		26.09	26.60	27.06	26.87	24.78
0435		25.30	25.83	26.40	26.53	24.84

(a) Plot on a graph temperature waves as functions of time and depth.

(b) Plot on a graph the vertical soil temperature profiles at 0435, 0835, 1235, 1635, 2035, and 0035 hr.

(c) Determine the damping depth and thermal diffusivity of the soil from the observed amplitudes, as well as from the times of temperature maxima (taken from the smoothed temperature waves) as functions of depth.

(d) Estimate the amplitude of the surface temperature wave and the time of maximum surface temperature from extrapolation of the soil temperature data.

(e) Estimating the temperature gradients near the surface from the vertical temperature profiles, determine the ground heat flux at 0435, 0835, 1235, 1635, 2035, and 0035 hr, if the heat capacity of the soil is 1.33×10^6 J m^{-3} K^{-1}.

8. From the soil temperature data given in the above problem, calculate the ground heat flux at 0435 and 1635 hr, using Eq. (4.4) with the measured soil heat capacity of 1.33×10^6 J m^{-3} K^{-1} and the average value of thermal diffusivity determined in Problem 7(c) above.

Chapter 5 | Air Temperature and Humidity in the PBL

5.1 FACTORS INFLUENCING AIR TEMPERATURE AND HUMIDITY

The following factors and processes influence the vertical distribution (profile) of air temperature in the atmospheric boundary layer:

- Type of air mass and its temperature just above the PBL, which depend on the synoptic situation and the large-scale circulation pattern
- Thermal characteristics of the surface and submedium, which influence the diurnal range of surface temperatures
- Net radiation at the surface and its variation with height, which determine the radiative warming or cooling of the surface and the PBL
- Sensible heat flux at the surface and its variation with height, which determine the rate of warming or cooling of the air due to convergence or divergence of sensible heat flux
- Latent heat exchanges during evaporation and condensation processes at the surface and in air, which influence the surface and air temperatures, respectively
- Warm or cold air advection as a function of height in the PBL
- Height of the PBL to which turbulent exchanges of heat are confined

Similarly, the factors influencing the specific humidity or mixing ratio of water vapor in the PBL are as follows:

- Specific humidity of air mass just above the PBL
- Type of surface, its temperature, and availability of moisture for evaporation and/or transpiration
- The rate of evapotranspiration or condensation at the surface and the variation of water vapor flux with height in the PBL
- Horizontal advection of water vapor as a function of height

- Mean vertical motion in the PBL and possible cloud formation and precipitation processes
- The PBL depth through which water vapor is mixed

The vertical exchanges of heat and water vapor are primarily through turbulent motions in the PBL; the molecular exchanges are important only in a very thin (of the order of 1 mm or less) layer adjacent to the surface. In the atmospheric PBL, turbulence is generated mechanically (due to surface friction and wind shear) and/or convectively (due to surface heating and buoyancy). These two types of motions are called forced convection and free convection, respectively. The combination of the two is sometimes referred to as mixed convection, although more often the designation ''forced'' or ''free'' is used to describe the same, depending on whether a mechanical or convective exchange process dominates.

The convergence or divergence of sensible heat flux leads to warming or cooling of air, similar to that due to net radiative flux divergence or convergence. It is easy to show that a gradient of 1 W m^{-3} in radiative or sensible heat flux will produce a change in air temperature at the rate of about 3°C hr^{-1}. Similarly, divergence or convergence of water vapor flux leads to decrease or increase of specific humidity with time.

Horizontal advections of heat and moisture become important only when there are sharp changes in the surface characteristics (e.g., land to water, rural to urban, and vice versa) or in air-mass characteristics (e.g., during the passage of a front) in the horizontal. Mean vertical motions forced by changes in topography often lead to rapid changes in temperature and humidity of air, as well as to local cloud formation and precipitation processes, as moist air rises over the mountain slopes. Equally sharp changes occur on the lee sides of the mountains. Even over a relatively flat terrain, the subsidence motion often leads to considerable warming and drying of the air in the PBL.

5.2 BASIC THERMODYNAMIC RELATIONS AND DEFINITIONS

The pressure in the lower atmosphere decreases with height in accordance with the hydrostatic equation

$$\partial P/\partial z = -\rho g \qquad (5.1)$$

Another fundamental relationship between the commonly used thermodynamic variables is the equation of state or the ideal gas law

$$P = R\rho T = (R_*/m)\rho T \qquad (5.2)$$

in which R is the specific gas constant, R_* is the absolute gas constant, and m is the mean molecular mass of air.

If a parcel of air is lifted upward in the atmosphere, its pressure will decrease in response to decreasing pressure of its surroundings. This will lead to a decrease in the temperature and, possibly, to a decrease in the density of the parcel (due to expansion of the parcel) in accordance with Eq. (5.2) and the first law of thermodynamics, which yields the relation

$$dH = \rho c_p \, dT - dP \qquad (5.3)$$

where dH is the heat added to the parcel per unit volume. If there is no exchange of heat between the parcel and the surrounding environment, the above process is called adiabatic. Vertical turbulent motions in the PBL, which carry air parcels up and down, are rapid enough to justify the assumption that changes in thermodynamic properties of such parcels are adiabatic ($dH = 0$). The rate of change of temperature of such a parcel with height can be calculated from Eqs. (5.1) and (5.3) and is characterized by the adiabatic lapse rate (Γ)

$$\Gamma \equiv -(\partial T/\partial z)_{\text{ad}} = g/c_p \qquad (5.4)$$

This is also the temperature lapse rate (rate of decrease of temperature with height) in an adiabatic atmosphere, which under dry or unsaturated conditions amounts to 0.0098 K m^{-1} or nearly 10 K km^{-1}.

The relationship between the changes of temperature and pressure in a parcel moving adiabatically, as well as those in an adiabatic atmosphere, are also given by Eqs. (5.2) and (5.3), with $dH = 0$.

$$dT/T = (R/c_p)(dP/P) \qquad (5.5)$$

integration of which gives the Poisson equation

$$T = T_0(P/P_0)^{R/c_p} \qquad (5.6)$$

where T_0 is the temperature corresponding to the reference pressure P_0.

Equation (5.6) is used to relate the potential temperature Θ, defined as the temperature which an air parcel would have if it were brought down to a pressure of 1000 mbar adiabatically from its initial state, to the actual temperature T as

$$\Theta = T(1000/P)^{R/c_p} \qquad (5.7)$$

where P is in millibars. The potential temperature has the convenient property of being conserved during vertical movements of an air parcel, provided heat is not added or removed during such excursions. Then, the parcel may be identified or labeled by its potential temperature. In an adiabatic atmosphere, potential temperature remains constant with height. For a nonadiabatic or diabatic atmosphere, it is easy to show from Eq. (5.7) that, to a good approximation,

$$\frac{\partial \Theta}{\partial z} = \frac{\Theta}{T} \left(\frac{\partial T}{\partial z} + \Gamma \right) \simeq \frac{\partial T}{\partial z} + \Gamma \qquad (5.8)$$

The approximation $\Theta/T \simeq 1$ is particularly useful in the surface layer, where the above ratio may not deviate from unity by more than 10%.

The above relations are applicable to both dry and moist atmospheres, so long as water vapor does not condense and one uses the appropriate values of thermodynamic properties for the moist air. In practice, it is found to be more convenient to use constant dry-air properties such as R and c_p, but to modify some of the relations or variables for the moist air.

5.2.1   MOIST UNSATURATED AIR

There are more than a dozen variables used by meteorologists that directly or indirectly express the moisture content of the air (see, e.g., Hess, 1959, Chap. 4). Of these the most often used in micrometeorology is the specific humidity (Q), defined as the ratio, $M_w/(M_w + M_d)$, of the mass of water vapor to the mass of moist air containing the water vapor. In magnitude, it does not differ significantly from the mixing ratio, defined as the ratio, M_w/M_d, of the mass of water vapor to the mass of dry air containing the vapor. Both are directly related to the water vapor pressure (e), which is the partial pressure exerted by water vapor and is a tiny fraction of the total pressure (P) anywhere in the PBL. To a good approximation,

$$Q \simeq m_w e/m_d P = 0.622 e/P \qquad (5.9)$$

where m_w and m_d are the mean molecular masses of water vapor and dry air, respectively.

The water vapor pressure is always less than the saturation vapor pressure (e_s), which is related to the temperature through the Clausius–Clapeyron equation (Hess, 1959, Chap. 4)

$$\frac{de_s}{e_s} = \frac{m_w L_e}{R_*} \frac{dT}{T^2} \qquad (5.10)$$

or in the integral form, after considering the fact that at $T = 273$ K, $e_s = 6.11$ mbar

$$\ln \frac{e_s}{6.11} = \frac{m_w L_e}{R_*} \left(\frac{1}{273} - \frac{1}{T} \right) \qquad (5.11)$$

in which e_s is in millibars.

$$\frac{PQ}{0.622} = 0.622 \, \frac{e}{P} \qquad KPA \cdot \frac{9}{kg}$$

Using Dalton's law of partial pressures ($P = P_d + e$), the equation of state for the moist air can be written as

$$P = \frac{R_*}{m} \rho T = \frac{R_*}{m_d} \frac{M_d}{V} T + \frac{R_*}{m_w} \frac{M_w}{V} T$$

or, after some algebra,

$$P = \frac{R_*}{m_d} \rho T \left[1 + \left(\frac{m_d}{m_w} - 1 \right) Q \right] \qquad (5.12)$$

After comparing Eq. (5.12) to Eq. (5.2) it becomes obvious that the factor in the brackets on the right-hand side of Eq. (5.12) is the correction to be applied to the specific gas constant for dry air, R_d, in order to obtain the specific gas constant for the moist air. Instead of taking R as a humidity-dependent variable, in practice, the above correction is applied to the temperature in defining the virtual temperature

$$T_v = T \left[1 + \left(\frac{m_d}{m_w} - 1 \right) Q \right] \simeq T(1 + 0.61Q) \qquad (5.13)$$

so that the equation of state for the moist air is given by

$$P = (R_*/m_d)\rho T_v = R\rho T_v \qquad (5.14)$$

The virtual temperature is obviously the temperature which dry air would have if its pressure and density were equal to those of the moist air. Note that the virtual temperature is always greater than the actual temperature and the difference between the two, $T_v - T \simeq 0.61 \, QT$, is never more than 7 K, which may occur over warm tropical oceans, and is usually less than 2 K.

For moist air one can define the virtual potential temperature, Θ_v, similar to that for dry air and use Eqs. (5.7) and (5.8) with Θ and T replaced by Θ_v and T_v, respectively.

The mixing ratio of water vapor in air being always very small (<0.03), the specific heat capacity is not significantly different from that of dry air. Therefore, changes in state variables of moving parcels in an unsaturated atmosphere are not significantly different from those in a dry atmosphere, so long as the processes remain adiabatic.

5.2.2 SATURATED AIR

In saturated air, a part of the water vapor may condense and, consequently, the latent heat of condensation may be released. If condensation products (e.g., water droplets and ice crystals) remain suspended in an upward-moving parcel of saturated air, the saturated lapse rate, or the rate at which the saturated air cools as it expands with the increase in height, must be lower than the dry adiabatic lapse rate Γ, because of the addition of the latent heat of condensation in the former. If there is no exchange of heat between the parcel and the surrounding environment, such a process is called moist adiabatic.

From the first law of thermodynamics for the moist adiabatic process and the equation of state, the moist adiabatic or saturated lapse rate is given by (Hess, 1959):

$$\Gamma_s = - \left(\frac{\partial T}{\partial z} \right)_s = g \left(c_p + L_e \frac{dQ_s}{dT} \right)^{-1} \qquad (5.15)$$

where dQ_s/dT is the slope of the saturation specific humidity (Q_s) versus temperature curve, as given by Eqs. (5.9) and (5.10). Unlike the dry adiabatic lapse rate, the saturated adiabatic lapse rate is quite sensitive to temperature. For example, at $T = 273$ K, $\Gamma_s = 0.0069$ K m^{-1} and at $T = 303$ K, $T_s = 0.0036$ K m^{-1}, both at the standard sea-level pressure of 1000 mbar.

In the real atmospheric situation, some of the condensation products are likely to fall out of an upward moving parcel, while others may remain suspended as cloud particles in the parcel. As a result of these changes in mass and composition of the parcel, some exchange of heat with the surrounding air must occur. Therefore the processes within the parcel are not truly adiabatic, but are called pseudoadiabatic. The mass of condensation products being only a tiny fraction of the total mass of parcel, however, the pseudoadiabatic lapse rate is not much different from the moist adiabatic lapse rate.

The difference between the dry adiabatic and moist adiabatic lapse rates accounts for some interesting orographic phenomena, such as the formation of clouds and possible precipitation on the upwind slopes of mountains and warm and dry downslope winds (e.g., Chinook and Foen winds) near the lee side base (see e.g., Rosenberg et al., 1983, Chap. 3).

5.3   STATIC STABILITY

The variations of temperature and humidity with height in the PBL lead to density stratification with the consequence that an upward- or downward-moving parcel of air will find itself in an environment whose density will, in general, differ from that of the parcel, after accounting for the adiabatic cooling or warming of the parcel. In the presence of gravity this density difference must lead to the application of a buoyancy force on the parcel which would accelerate or decelerate its vertical movement. If the vertical motion of the parcel is enhanced, i.e., the parcel is accelerated by the buoyancy force, the environment is called statically unstable. On the other hand, if the parcel is decelerated, the atmosphere is called stable or stably stratified. When the atmosphere exerts no buoyancy force on the parcel at all, it is considered neutral. In general the buoyancy force or acceleration exerted on the parcel may vary with height and so does the static stability.

One can derive an expression for the buoyant acceleration, a_b, on a parcel of air by using the Archimedes principle as

$$a_b = g \left(\frac{\rho - \rho_p}{\rho_p} \right) \tag{5.15}$$

in which the subscript p refers to the parcel. Further, using the equation of state for the moist air, Eq. (5.15) can also be written as

$$a_b = -g \left(\frac{T_v - T_{vp}}{T_v} \right) \tag{5.16}$$

or, in terms of the gradient of virtual temperature or potential temperature and a small displacement, Δz, from the equilibrium position

$$a_b \simeq -\frac{g}{T_v} \left(\frac{\partial T_v}{\partial z} + \Gamma \right) \Delta z = -\frac{g}{T_v} \frac{\partial \Theta_v}{\partial z} \Delta z \tag{5.17}$$

Equation (5.17) gives a criterion, as well as a quantitative measure, of static stability of the atmosphere as a function of $\partial T_v / \partial z$ or $\partial \Theta_v / \partial z$. In particular, the static stability parameter is defined as

$$s = (g/T_v)(\partial \Theta_v / \partial z) \tag{5.18}$$

Thus, qualitatively, the atmospheric stability can be divided into three broad categories:

1. Unstable, when $s < 0$, $\partial \Theta_v / \partial z < 0$, or $\partial T_v / \partial z < -\Gamma$
2. Neutral, when $s = 0$, $\partial \Theta_v / \partial z = 0$, or $\partial T_v / \partial z = -\Gamma$
3. Stable, when $s > 0$, $\partial \Theta_v / \partial z > 0$, or $\partial T_v / \partial z > -\Gamma$

The magnitude of s provides a quantitative measure of static stability. It is quite obvious from Eq. (5.17) that vertical motions are generally enhanced ($a_b > 0$) in an unstable atmosphere, while they are suppressed ($a_b < 0$) in a stably stratified environment.

On the basis of virtual temperature gradient or lapse rate, relative to the adiabatic lapse rate Γ, an atmospheric layer can be variously character-ized as follows:

1. Superadiabatic, when $\partial T_v/\partial z < -\Gamma$
2. Adiabatic, when $\partial T_v/\partial z = -\Gamma$
3. Subadiabatic, when $0 > \partial T_v/\partial z > -\Gamma$
4. Isothermal, when $\partial T_v/\partial z = 0$
5. Inversion, when $\partial T_v/\partial z > 0$

Figure 5.1 gives a schematic of these stability categories in the lower part of the PBL. Here, the surface temperature is assumed to be the same and virtual temperature profiles are assumed to be linear for convenience only; actual profiles are usually curvilinear, as will be shown later.

Sometimes, characterizations such as "extreme," "moderate," and "slight" are used to indicate the degree of stability or instability of the atmosphere. Including the neutral or near-neutral category, this divides atmospheric stability into six or seven categories, which have been desig-nated in the air pollution literature by certain letters (e.g., Pasquill's A–F stability categories) or numbers (e.g., D. B. Turner's categories 1–7). These are usually defined on the basis of wind speed at a 10-m height, strength of insolation, and cloudiness, as shown in Table 5.1.

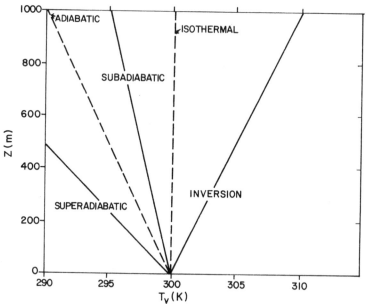

Fig. 5.1 Schematic of the various stability categories on the basis of virtual temperature gradient.

Table 5.1

Meteorological Conditions Defining Pasquill's Stability Categories[a]

Surface wind speed (m sec^{-1})	Daytime insolation			Nighttime	
	Strong	Moderate	Slight	Cloudiness $\geq$ 4/8	Cloudiness $\leq$ 3/8
<2	A	A–B	B	—	—
2	A–B	B	C	E	F
4	B	B–C	C	D	E
6	C	C–D	D	D	D
>6	C	D	D	D	D

[a] A, Extremely unstable; B, moderately unstable; C, slightly unstable; D, near-neutral (applicable to heavy overcast day or night); E, slightly stable; F, moderately stable.

5.4 MIXED LAYERS AND INVERSIONS

In moderate to extremely unstable conditions, typically encountered during midday and afternoon periods over land, mixing within bulk of the boundary layer is intense enough to make conservable scalars, such as Θ and Θ_v, distributed uniformly, independent of height. The layer over which this occurs is called the mixed layer, which is an adiabatic layer overlying a superadiabatic surface layer. To a lesser extent, the specific humidity, momentum, and concentrations of pollutants from distant sources also have uniform distributions in the mixed layer. Mixed layers are found to occur most persistently over the tropical and subtropical oceans.

Sometimes, the term "mixing layer" is used to designate a layer of the atmosphere in which there is significant mixing, even though the conservable properties may not be uniformly mixed. In this sense, the turbulent PBL is also a mixing layer, a large part of which may become well mixed (mixed layer) during unstable conditions. Mixing layers are also found to occur sporadically or intermittently in the free atmosphere above the PBL, whenever there are suitable conditions for breaking of internal gravity waves and generation of turbulence in an otherwise smooth, streamlined flow.

The term "inversion" in common usage applies to an atmospheric condition when air temperature increases with height. Micrometeorologists sometimes refer to inversion as any stable atmospheric layer in which potential temperature (more appropriately, Θ_v) increases with height. The strength of inversion is expressed in terms of the layer-averaged $\partial \Theta_v / \partial z$ or the difference $\Delta \Theta_v$ across the given depth of the inversion layer.

Inversion layers are variously classified by meteorologists according to their location (e.g., surface and elevated inversions), the time (e.g., nocturnal inversion), and the mechanism of formation (e.g., radiation, evaporation, advection, subsidence, and frontal and sea-breeze inversions). Because vertical motion and mixing are considerably inhibited in inversions, the knowledge of the frequency of occurrence of inversions and their physical characteristics (e.g., location, depth, and strength of inversion) is very important to air pollution meteorologists. Pollutants released within an inversion layer often travel long distances without much mixing and spreading. These are then fumigated to the ground level, as soon as the inversion breaks up to the level of the ribbonlike thin plume, resulting in large ground-level concentrations. Low-level and ground inversions act as lids, preventing the downward diffusion or spread of pollutants from elevated sources. Similarly, elevated inversions put an effective cap on the upward spreading of pollutants from low-level sources. An inversion layer usually caps the daytime unstable or convective boundary layer, confining the pollutants largely to the PBL. Consequently, the PBL depth is commonly referred to as the mixing depth in the air quality literature.

The most persistent and strongest surface inversions are found to occur over the Antarctica and Arctic regions; the tradewind inversions are probably the most persistent elevated inversions capping the PBL.

5.5   VERTICAL TEMPERATURE AND HUMIDITY PROFILES

A typical sequence of observed potential temperature profiles at 3-hr intervals during the course of a day is shown in Fig. 5.2. These were obtained from radiosonde measurements during the 1967 Wangara Experiment near Hay, New South Wales, Australia, on a day when there were clear skies, very little horizontal advections of heat and moisture, and no frontal activity within 1000 km.

Note that, before sunrise and at the time of the minimum in the near-surface temperature, the Θ profile is characterized by nocturnal inversion, which is produced as a result of radiative cooling of the surface. The nocturnal boundary layer (NBL) is stably stratified, in which the vertical turbulent exchanges are greatly suppressed. Soon after sunrise, the surface heating leads to an upward exchange of sensible heat and subsequent warming of the lowest layer due to heat flux convergence. This process progressively erodes the nocturnal inversion from below and replaces it with an unstable or convective boundary layer (CBL) whose depth h grows with time (see Fig. 5.2c). The rate of growth of h usually attains maximum a few hours after sunrise when all the surface-based inversion has been eroded and slows down considerably in the late afternoon hours.

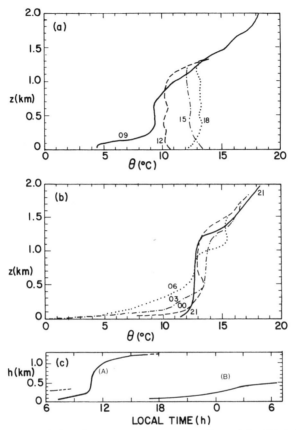

Fig. 5.2   Diurnal variation of potential temperature profiles and the PBL height during (a) day 33 and (b) days 33–34 of the Wangara Experiment. (c) Curve A, convective; Curve B, stable. [After Deardorff (1978).]

The potential temperature remains more or less uniform in the so-called mixed layer which comprises the bulk of the CBL. This layer is dominated by buoyance-generated turbulence, except during morning and evening transition hours when shear effects become important.

Just before sunset, there is net radiative loss of energy from the surface and, consequently, an inversion forms at the surface. The nocturnal inversion deepens in the early evening hours and sometimes throughout the night, as a result of radiative and sensible heat flux divergence. Still, the remnants of the previous afternoon's mixed layer may remain above the nocturnal inversion, as seen in Fig. 5.2b, although mixing processes in the former are likely to be very weak, due to the cutoff of heat energy from the surface.

Observed vertical profiles of specific humidity for the same Wangara day 33 are shown in Fig. 5.3. The surface during this period of drought was rather dry, with little or no growth of vegetation (predominantly dry grass, legume, and cottonbush). In the absence of any significant evapotranspiration from the surface, the specific humidity profiles are nearly uniform in the daytime PBL, with the value of Q changing largely in response to the evolution of the mixed layer.

In other situations where there is dry air aloft, but plenty of moisture available for evaporation at the surface, the evolution of Q profiles might be quite different from those shown in Fig. 5.3. In response to the increase in the rate of evaporation or evapotranspiration during the morning and early afternoon hours, Q may actually increase and attain its maximum value sometime in the afternoon and, then, decrease thereafter, as the evaporation rate decreases.

The evaporation from the surface, although strongest by day, may continue at a reduced rate throughout the night. Under certain conditions, when winds are light and the surface temperature falls below the dew point temperature of the moist air near the surface, the water vapor is likely to condense and deposit on the surface in the form of dew. The resulting inversion in Q profile leads to a downward flux of water vapor and further deposition.

Temperature and humidity profiles over open oceans do not show any significant diurnal variation when large-scale weather conditions remain unchanged, because the sea-surface temperature changes very little (typi-

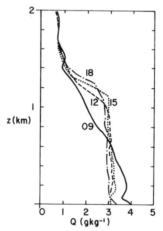

Fig. 5.3 Diurnal variation of specific humidity profiles during day 33 of the Wangara Experiment. [From Andre *et al.* (1978).]

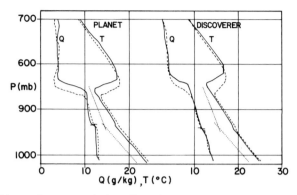

Fig. 5.4 Observed mean vertical profiles of temperature and specific humidity at two research vessels during ATEX. First ("undisturbed") period, ———; second ("disturbed") period, ---. ·····, Dry adiabatic and saturated adiabatic lapse rates. [After Augstein *et al.* Copyright © (1974) by D. Reidel Publishing Company. Reprinted by permission.]

cally, less than 0.5°C), if at all, between day and night. Figure 5.4 shows the observed profiles of T and Q from two research vessels (with a separation distance of about 750 km) during the 1969 Atlantic Tradewind Experiment (ATEX). Here, only the averaged profiles for the two periods designated as "undisturbed" (February 7–12) and "disturbed" (February 13–17) weather are given. During the undisturbed period, the temperature and humidity structure of the PBL is characterized by a mixed layer (depth ≃ 600 m), followed by an isothermal transition layer (depth ≃ 50 m) and a deep cloud layer extending to the base of the trade wind inversion. The same features can also be seen during the disturbed period, but with larger spatial differences.

On the other extreme, Fig. 5.5 shows the measured mean temperature profiles in the surface inversion layer at the Plateau Station, Antarctica (79°15′S, 40°30′E), in three different seasons (periods) and for different stability classes [identified here as (1) through (8) in the order of increasing stability and defined from both the temperature and wind data]. These include some of the most stable conditions ever monitored near the surface.

5.6 DIURNAL VARIATIONS

As a result of the diurnal variations of net radiation and other energy fluxes at the surface, there are large diurnal variations in air temperature and specific humidity in the PBL over land surfaces. The diurnal varia-

(a)

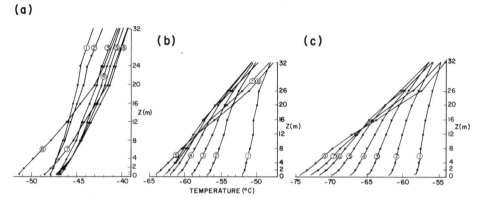

Fig. 5.5   Observed mean temperature profiles in the surface inversion layer at Plateau Station, Antarctica, for three different periods grouped under different stability classes. (a) Sunlight periods 1967, (b) transitional periods 1967, (c) dark season 1967. [After Lettau *et al.* (1977).]

tions of the same over large lakes and oceans are much smaller and often negligible.

The largest diurnal changes in air temperature are experienced over desert areas where strong large-scale subsidence limits the growth of the PBL and the air is rapidly heated during the daytime and is rapidly cooled at night. The mean diurnal range of screen-height temperatures is typically 20°C, with mean maximum temperatures of about 45°C during the summer months (Deacon, 1969).

The diurnal range of temperatures is reduced by the presence of vegetation and the availability of moisture for evaporation. Even more dramatic reductions occur in the presence of cloud cover, smoke, haze, and strong winds. The diurnal range of temperatures decreases rapidly with height in the PBL and virtually disappears at the maximum height of the PBL or inversion base, which is reached in the late afternoon. Some examples of observed diurnal variations of air temperature at different heights and under different cloud covers and wind conditions are given in Fig. 5.6.

The diurnal variation of specific humidity depends on the diurnal course of evapotranspiration and condensation, surface temperatures, mean winds, turbulence, and the PBL height. Large diurnal changes in surface temperature generally lead to large variations of specific humidity, due to the intimate relationship between the saturation vapor pressure and temperature. Figure 5.7 illustrates the diurnal course of water vapor pressure at three different heights. Here the secondary minimum of vapor pressure in the afternoon is believed to be caused by further deepening of the mixed layer after evapotranspiration reached its maximum value. This phenomenon may be even more marked at other places and times.

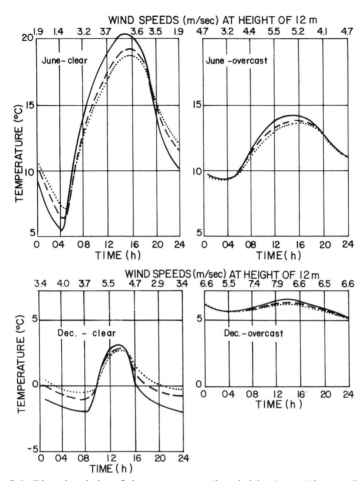

Fig. 5.6 Diurnal variation of air temperature at three heights (——, 1.2 m; ---, 7 m; ·····, 17 m) over downland in southern England for the various combinations of clear and overcast days in June and December. Wind speeds at a 12-m height are also indicated. [From Deacon (1969); after Johnson (1929).]

5.7 APPLICATIONS

The knowledge of air temperature and humidity in the PBL is useful and may have applications in the following areas:

- Determining the static stability of the lower atmosphere and identifying mixed layers and inversions
- Determining the PBL height, as well as the depth and strength of surface or elevated inversion

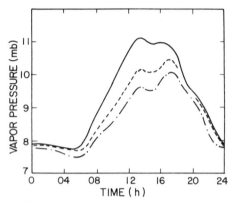

Fig. 5.7   Diurnal variation of water vapor pressure at three heights at Quickborn, Federal Republic of Germany, on clear May days. ——, 2 m; ---, 13 m; ·—·, 70 m. [From Deacon (1969).]

- Predicting air temperature and humidity near the ground and possible fog and frost conditions
- Determining the sensible heat flux and the rate of evaporation from the surface
- Calculating atmospheric radiation
- Estimating diffusion of pollutants from surface and elevated sources and possible fumigation
- Designing heating and air-conditioning systems

PROBLEMS AND EXERCISES

1. How do the mechanisms of heat transfer in the atmosphere differ from those in a subsurface medium, such as soil and rock?
2. Derive the following equation for the variation of air temperature with time, in the absence of any horizontal and vertical advections of heat:

$$\rho c_p (\partial T/\partial t) = \partial R_N/\partial z - \partial H/\partial z$$

Indicate how the above equation may be used to estimate the sensible heat flux at the surface, H_0, during the daytime.
3. (a) Derive an expression for the adiabatic lapse rate in a moist, unsaturated atmosphere and discuss the effect of moisture on the same.
   (b) Explain why the saturated adiabatic lapse rate must be smaller than the dry diabatic lapse rate.

4. (a) Derive the equation of state for the moist air and hence define the virtual temperature T_v.

   (b) Show that, to a good approximation,

   $$\partial T_v/\partial z = \partial T/\partial z + 0.61T(\partial Q/\partial z)$$

5. (a) Show that the buoyant acceleration on a parcel of air in a thermally stratified environment is given by

   $$a_b = -(g/T_v)(T_v - T_{vp})$$

   (b) Considering a small vertical displacement, Δz, from the equilibrium position of the parcel, show that

   $$a_b \simeq -(g/T_v)(\partial \Theta_v/\partial z)\Delta z$$

6. A moist air parcel at a temperature of 20°C and specific humidity of 10 g kg^{-1} is lifted adiabatically from the upwind base of a mountain, where the pressure is 1000 mbar, to the top, at 3000 m above the base, and is then brought down to the base on the other side of the mountain. Using appropriate equations or tables, calculate the following parameters:

   (a) The height above the base where air parcel would become saturated and the saturated adiabatic lapse rate at this level

   (b) The temperature of the parcel at the top of the mountain

   (c) The temperature of the parcel at the downwind base, assuming that the parcel remained unsaturated during its entire descent

7. The following measurements were made from the research vessel FLIP during the 1969 Barbados Oceanographic and Meteorological Experiment (BOMEX):

Height (m)	Wind speed (m sec^{-1})	Temperature (°C)	Specific humidity (g kg^{-1})
2.1	4.33	27.74	18.02
3.6	4.52	27.72	17.83
6.4	4.64	27.69	17.54
11.4	4.85	27.62	17.29

   (a) Calculate the virtual air temperature at each level.

   (b) Calculate $\partial T/\partial z$, $\partial \Theta/\partial z$, $\partial T_v/\partial z$, and $\partial \Theta_v/\partial z$ between the adjacent measurement levels.

   (c) Assess the percentage contribution of specific humidity gradient to the static stability, s, between the adjacent levels.

Chapter 6 | Wind Distribution in the PBL

6.1 FACTORS INFLUENCING WIND DISTRIBUTION

The magnitude and direction of near-surface winds and their variations with height in the PBL are of considerable interest to micrometeorologists. The following factors influence the wind distribution in the PBL:

- Large-scale horizontal pressure and temperature gradients in the lower atmosphere, which drive the PBL flow
- The surface roughness characteristics, which determine the surface drag and momentum exchange in the lower part of the PBL
- The earth's rotation, which makes wind turn with height
- The diurnal cycle of heating and cooling of the surface, which determines the thermal stratification of the PBL
- The PBL depth, which determines wind shears in the PBL
- Entrainment of the free atmospheric air into the PBL, which determines the momentum, heat, and moisture exchanges at the top of the PBL, as well as the PBL height
- Horizontal advections of momentum and heat, which affect both the wind and temperature profiles in the PBL
- Large-scale horizontal convergence or divergence and the resulting mean vertical motion at the top of the PBL
- Presence of clouds and precipitation in the PBL, which influence its thermal stratification
- Surface topographical features, which give rise to local or mesoscale circulations

Some of these factors will be discussed in more detail in the following text, while others are considered outside the scope of this introductory text.

67

6.2 GEOSTROPHIC AND THERMAL WINDS

The PBL is essentially driven by large-scale atmospheric motions which are set up in response to spatial variations of air pressure and temperature. The geostrophic winds are related to or defined in terms of the pressure gradients as

$$U_g = -\frac{1}{\rho f}\frac{\partial P}{\partial y}; \qquad V_g = \frac{1}{\rho f}\frac{\partial P}{\partial x} \tag{6.1}$$

in which U_g and V_g are the components of geostrophic wind vector **G** in the x and y directions, respectively, and f is the Coriolis parameter, which is related to the rotational speed of the earth (Ω) and latitude (ϕ) as

$$f = 2\Omega \sin \phi \tag{6.2}$$

The geostrophic winds are the winds which would occur as a result of the simple geostrophic balance between the pressure gradient and Coriolis (rotational) forces in a frictionless atmosphere with no advective and local accelerations. Equation (6.1) implies that **G** must be parallel to the isobars with low pressure to the left in the Northern Hemisphere ($f > 0$) (see Fig. 6.1) and low pressure to the right in the Southern Hemisphere ($f < 0$). The absence of advective accelerations further implies that, locally, isobars are equispaced, parallel, straight lines. The geostrophic balance is often assumed to occur, i.e., $U = U_g$ and $V = V_g$, at the top of and outside the PBL. Within the PBL, however, actual winds differ from the

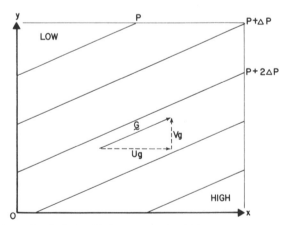

Fig. 6.1 Schematic of relationship between geostrophic winds and isobars for the Northern Hemisphere.

geostrophic winds because of the surface friction and the vertical exchange of momentum.

The horizontal pressure gradients or geostrophic winds may vary with height in response to horizontal temperature gradients. The vertical gradients of geostrophic winds, i.e., geostrophic wind shears, are given by the thermal wind equations

$$\frac{\partial U_g}{\partial z} = -\frac{g}{fT}\frac{\partial T}{\partial y} + \frac{U_g}{T}\frac{\partial T}{\partial z} \simeq -\frac{g}{fT}\frac{\partial T}{\partial y}$$

$$\frac{\partial V_g}{\partial z} = \frac{g}{fT}\frac{\partial T}{\partial x} + \frac{V_g}{T}\frac{\partial T}{\partial z} \simeq \frac{g}{fT}\frac{\partial T}{\partial x}$$

(6.3a)

which can be derived from Eq. (6.1) in conjunction with the hydrostatic equation and the equation of state (see, e.g., Hess, 1959, Chapter 12). The approximations on the right-hand side of Eq. (6.3a) ignore the stability-dependent terms, which may account for no more than 10% variation in geostrophic winds per kilometer of height. The approximate thermal wind equations imply that the geostrophic wind shear vector $\partial \mathbf{G}/\partial z$ must be parallel to the isotherms with colder air to the left in the Northern Hemisphere (see, e.g., Fig. 6.2). A part of the geostrophic wind shear is due to the normal (climatological) decrease of temperature in going toward the poles, while a substantial part may be due to local temperature gradients created by surface topographical features (e.g., land–sea, urban–rural, and mountain–valley contrasts) and/or the synoptic weather situation. Note that a horizontal temperature gradient of only 1°C per 100 km will

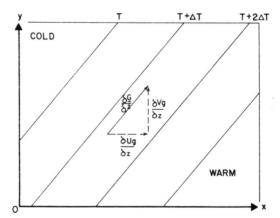

Fig. 6.2 Schematic of relationship between thermal winds or geostrophic shears and isotherms for the Northern Hemisphere.

cause a geostrophic wind shear of about 3.3 m sec^{-1} km^{-1} or 0.0033 sec^{-1} in middle latitudes.

The changes in geostrophic wind components over a given layer depth, Δz, can be calculated from the finite-difference form of Eq. (6.3a):

$$\Delta U_g = -\frac{\Delta z g}{fT}\frac{\partial T}{\partial y}; \qquad \Delta V_g = \frac{\Delta z g}{fT}\frac{\partial T}{\partial x} \qquad (6.3b)$$

Horizontal temperature gradients or geostrophic shears are generally assumed to be constant (independent of height) in the PBL. It follows from Eq. (6.3a) or (6.3b) that geostrophic winds must vary with height in the presence of finite-temperature gradients. Figure 6.3 illustrates for the Northern Hemisphere the relationship between the geostrophic winds at

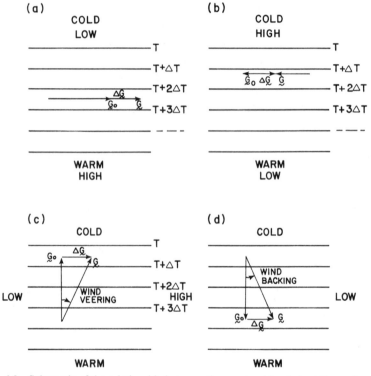

Fig. 6.3 Schematic of the relationship between the geostrophic winds at the surface and the top of the PBL for different orientations of surface isobars or G_0 and surface isotherms: (a) parallel geostrophic flow around a cold low or a warm high, (b) parallel flow around a cold high or a warm low, (c) veering geostrophic flow in warm advection, (d) backing flow in cold advection.

the surface and the top of the PBL (taking $\Delta z = h$) for different orientations of the surface geostrophic winds G_0 with respect to the isotherms. Again, isobars and isotherms are taken as parallel straight lines for the ideal case of no advective acceleration.

Note that when G_0 is parallel to the isotherms (no temperature advection), the effect of the horizontal temperature gradient is only to increase or decrease the magnitude of geostrophic winds, without causing any change in the direction. This will be the rare situation when isobars are parallel to the isotherms throughout the layer. In general, however, the surface isobars are not parallel to the isotherms and the surface geostrophic wind has a component from cold toward warm temperatures or vice versa. Such a flow tends to change the vertical temperature distribution due to horizontal advection of cold or warm air. Figure 6.3 shows that both the magnitude and the direction of the geostrophic wind may change with height in the presence of heat advection. The geostrophic wind turns cyclonically (backs) with height in the case of cold air advection, and it turns anticyclonically (veers) with height in the case of warm advection. Actual winds within the PBL may also be expected to back or veer with height in response to the backing and veering of geostrophic winds.

The atmosphere is called barotropic, when there are no geostrophic shears or horizontal temperature gradients, and baroclinic, when there are significant geostrophic shears. In the same way, atmospheric boundary layers may be classified as barotropic and baroclinic PBLs. The term baroclinicity (or baroclinity) is often used as a synonym for geostrophic wind shear. It is not difficult to see that the presence of geostrophic shears or baroclinity in the PBL is more a rule than the exception.

6.3   THE EFFECTS OF FRICTION

Because air is a viscous fluid, its velocity relative to the earth's surface must vanish right at the surface. This does not happen abruptly, but the retarding influence of the surface on atmospheric winds pervades the whole depth of the PBL. Consequently, the wind speed decreases gradually and the mean momentum is transferred downward in going toward the surface. Turbulence provides an efficient mechanism for the transfer of momentum from one level to another in the vertical. Also, vertical exchange of horizontal momentum results in the so-called frictional force on any fluid element. Because of this, the simple geostrophic balance between the pressure gradient and Coriolis forces, which may exist outside the PBL, cannot be expected to occur within the PBL.

Figure 6.4 is a schematic of the balance of pressure gradient, Coriolis, and friction forces on fluid elements at different levels in a barotropic PBL. The corresponding geostrophic and actual wind vectors are also shown with the stipulation that the Coriolis force must always be normal to **V** and proportional to wind speed, while the friction force must be perpendicular to the ageostrophic wind vector (**V** − **G**). These conditions follow from the equation of mean motion in the absence of local and advective accelerations.

$$2\rho\mathbf{\Omega} \times (\mathbf{V} - \mathbf{G}) = \partial\tau/\partial z \qquad (6.4)$$

Note that in a rotating frame of reference or in the presence of directional shear, the friction force on a fluid element need not be parallel and opposite to the velocity vector, as commonly depicted in many textbook schematics of the force balance in the friction layer. Figure 6.4a shows that, right at the surface where the Coriolis force disappears, the friction

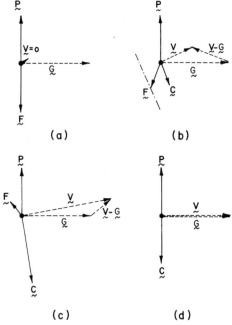

Fig. 6.4 Schematic of the balance of forces on a fluid parcel at different heights in a barotropic PBL: (a) at the surface, (b) in the surface layer, (c) in the middle of the PBL, (d) at the top of the PBL. Actual and geostrophic velocities are also indicated. [After Arya (1986).]

force must exactly balance the pressure gradient force and, hence, make a large angle with the direction of near-surface wind or surface shear. As wind speed increases with height, with little change in the wind direction through the surface layer, the balance of forces requires that the friction force decrease and rotate anticyclonically with increasing height (see Fig. 6.4b). The same tendency continues through the outer part of the PBL also, as wind veers and increases in magnitude (it often becomes super-geostrophic in the middle of the PBL), as shown in Fig. 6.4c. Near the top of the PBL, the friction force becomes small in magnitude and highly variable in direction, in response to the changes in the direction of $\mathbf{V} - \mathbf{G}$, as velocity oscillates about the geostrophic equilibrium. Ignoring any such oscillations, as well as any inertial oscillations that may be present in the real atmosphere, the force balance at the top of the PBL is depicted in Fig. 6.4d.

The above considerations of the balance of forces suggest that winds must veer with height in a barotropic PBL. The difference between the wind direction at some height and that near the surface is called the veering angle, which normally increases with height in a barotropic PBL. In the presence of thermal winds (geostrophic wind shear), however, actual winds may veer or back with increasing height, in response to the veering or backing of geostrophic winds and the frictional veering. In practice, frictional veering is determined as the difference between the actual wind veering and the geostrophic veering.

The angle between the directions of actual and geostrophic winds is called the cross-isobar angle (α) of the flow, which normally decreases with height and vanishes at the top of the PBL (assuming that geostrophic balance occurs there). The surface cross-isobar angle (α_0) which near-surface winds make with surface isobars is of considerable interest to meteorologists. It is found to depend on the surface roughness, latitude, and the surface geostrophic wind, as well as on the stability, baroclinity, and height of the PBL. For the neutral, barotropic PBL in middle and high latitudes, α_0 is found to range from about 10° for relatively smooth (e.g., water, ice, and snow) surfaces to about 35° for very rough (e.g., forests and urban areas) surfaces. It increases with increasing stability and de-creasing latitude, so that a wide range of $\alpha_0 \simeq 0\text{-}45°$ occur in the stratified barotropic PBL. In the presence of thermal winds (baroclinic PBL), the range of observed surface cross-isobar angles becomes wider (say, -20 to 70°). Under certain conditions of strong, warm air advection α_0 may even become negative, implying a component of the near-surface winds to be directed toward the surface "high" (see, e.g., Arya and Wyngaard, 1975; Arya, 1978).

6.4   THE EFFECTS OF STABILITY

We have discussed earlier how the diurnal cycle of heating and cooling of the surface, in conjunction with radiative, convective, and advective processes occurring within the PBL, determine the static or thermal stability of the PBL. Through its strong influence on the vertical movements of air parcels, thermal stability strongly influences the vertical exchange of momentum and, hence, the wind distribution in the PBL.

On a clear day the surface warms up relative to the air above, in response to solar heating. This gives rise to a variety of convective circulations, such as plumes, thermals, and, occasionally, dust devils, which can directly transfer momentum and heat in the vertical direction. The upward transfer of heat through the lower part of the PBL is largely responsible for convective or buoyancy-generated turbulence. The resulting vigorous mixing of momentum leads to considerable weakening and, sometimes, elimination of mean wind shears in the PBL. Thus, it is not uncommon to find nearly uniform wind speed and wind direction profiles through much of the daytime convective PBL. Strong wind shears are largely confined to the lower part of the surface layer and the shallow transition layer near the base of inversion which often caps the convective mixed layer. Changes in mean wind direction between the surface and the inversion base are typically less than 10°.

On a clear evening and night, on the other hand, the surface cools down in response to longwave radiation and a surface inversion begins to form and develop. Since buoyancy inhibits vertical momentum exchanges in the inversion layer, significant wind speed and direction shears develop in this layer. Wind direction changes of 30° or more across the shallow nocturnal boundary layer, which may comprise only a part of the surface inversion layer, are not uncommon. The wind speed profile is generally characterized by a low-level jet in which winds are often supergeostrophic. The winds in the outer part of the NBL may undergo inertial oscillations in response to the inertially oscillating geostrophic flow. Internal gravity waves may also develop in such a stratified environment; such waves frequently appear mixed with turbulence.

Thermal stability also has a strong influence on the PBL height, which, in turn, affects the wind distribution. In the morning hours following sunrise, the PBL grows rapidly at first, in response to heating from below. This growth continues throughout the day, although at a progressively decreasing rate, resulting into a 1- to 2-km-deep mixed layer by midafternoon. Immediately following the evening transition period, when the sensible heat flux at the surface changes sign, the convective PBL suddenly collapses and is replaced by a much shallower nocturnal boundary layer.

For a given wind speed at the top of the PBL, one would expect the average wind shear in the PBL to be an order of magnitude larger at night than that during the midday period, because of the difference in the PBL heights. Very close to the surface, however, wind shear values are generally larger during the daytime than they are at night.

In the presence of wind shear, a dynamic stability parameter such as the Richardson number

$$\text{Ri} = \frac{g}{T_v} \frac{\partial \Theta_v}{\partial z} \left| \frac{\partial \mathbf{V}}{\partial z} \right|^{-2} \tag{6.5}$$

is considered to be a more appropriate stability parameter than the static stability parameter $s = (g/T_v)(\partial \Theta_v/\partial z)$ introduced earlier. Ri is dimensionless and has the same sign as s. It will be shown in Chapter 8 that the Richardson number is a better measure of the intensity of mixing (turbulence) and provides a simple criterion for the existence or nonexistence of turbulence in a stably stratified environment (a large positive value of Ri > 0.25 is indicative of weak and decaying turbulence or a completely nonturbulent environment). Therefore, the vertical distribution of Ri may be used to determine the vertical extent of the boundary layer when other, more direct, measurements of the PBL height are not available.

6.5 OBSERVED WIND PROFILES

A considerable amount of wind profile data has been collected by meteorologists in the course of a number of large and small field experiments, as well as from routine upper air soundings. Observations are made using pilot balloons, rawindsondes, tethersondes, doppler radars, instrumented aircraft, and tall towers. Consequently, scientists have developed a good understanding of the effects of surface roughness and stability on wind distribution in the PBL and a fair understanding of the influence of baroclinity on the same. More complicated effects of entrainment, advection, complex topography, clouds, and precipitation have received little or no attention from boundary layer meteorologists and remain poorly understood. The underlying philosophy in micrometeorology has been to study simple situations first in order to have a better understanding of the basic phenomena and then include progressively more complicating factors. Unfortunately, difficulties of measuring and computing turbulence in the atmospheric boundary layer have restricted micrometeorologists for a long time to the study of the more or less idealized (horizontally homogeneous, stationary, nonentraining, dry, etc.) PBL. Only recently, some efforts have been made to study the effects of more complicating factors which influence the real-world PBL.

First, we present a few selected wind profiles for the "ideal" PBL, taken under different stability conditions, in order to illustrate some qualitative effects of stability. Figure 6.5 shows the pibal wind and potential temperature profiles under fairly convective conditions during day 33 of the Wangara Experiment in southern Australia. Here, U and V are the horizontal wind components in the x and y directions, respectively, with the x axis parallel to the direction of near-surface winds and the y axis normal to the same (this choice of coordinate axes is more commonly used in micrometeorology than the geographical coordinate system, which is used in other branches of meteorology).

Note that despite the large geostrophic shear present on this day, wind profiles show the characteristic features of nearly uniform distributions in the convective mixed layer (the undulations in the PBL are probably caused by large eddies and organized secondary motions whose effects are not smoothed out in pibal soundings), and strong gradients in the surface layer below and the transition layer above. Similar features have been observed in the convective boundary layers (CBLs) over other homogeneous land and ocean sites (see, e.g., Fig. 6.6a).

The observed mean wind and virtual potential temperature profiles under moderately unstable conditions in a trade wind marine PBL are shown in Fig. 6.6b. At the time of these observations widely scattered shallow convective clouds were present whose bases coincided with the inversion base. Note that although the Θ_v profile is uniform throughout the subcloud layer, the wind profiles show significant shears in this layer. Still, the total wind veering across the unstable PBL remains small (about 8°), due to the

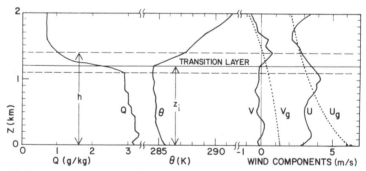

Fig. 6.5 Measured wind, potential temperature, and specific humidity profiles in the PBL under convective conditions on day 33 of the Wangara Experiment. [From Deardorff (1978).]

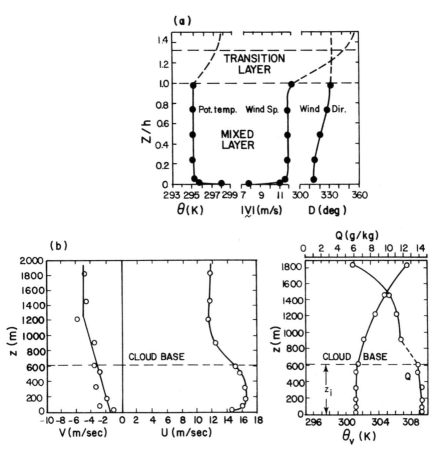

Fig. 6.6   Observed vertical profiles of mean wind components, or wind speed and direction, potential temperature, and specific humidity in the PBL. (a) Convective conditions overland. [After Kaimal *et al.* (1976).] (b) Unstable conditions over the ocean. [After Pennell and Lemone (1974).]

combined effects of mechanical and buoyant mixing, smooth ocean surface, and low latitude.

The theoretician's ideal of a steady state, neutral, barotropic PBL is so rare in the atmosphere that relevant observations of the same do not exist. The wind profiles taken under slightly unstable and slightly stable conditions during the morning and evening transition hours differ considerably, due to the effects of nonstationarity and even slight stability or instability. An average of profiles taken during completely overcast and very windy conditions might be more representative of a neutral PBL.

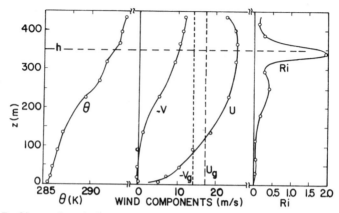

Fig. 6.7 Observed vertical profiles of mean wind components and potential temperature and the calculated Ri profile in the nocturnal PBL under moderately stable conditions. [From Deardorff (1978); after Izumi and Barad (1963).]

From observations of a stably stratified nocturnal boundary layer, one can perhaps distinguish between the two broad stability regimes: (1) the moderately stable regime in which turbulent exchanges are more or less continuous in time and space through at least the lower half of the PBL, and (2) the very stable regime in which turbulent exchanges occur only intermittently in time and space through most of the PBL. The former can exist over land at night only during strong winds, and more likely at sea when the air is slightly warmer than the sea surface. The typical profiles of wind, potential temperature, and Richardson number in this moderately stable NBL are shown in Fig. 6.7. These are based on the measurements from a tall tower near Dallas, Texas. Note that the U profile shows a pronounced maximum. The nose of the low-level jet often coincides with the top of the nocturnal PBL, since it also represents the level of maxi-

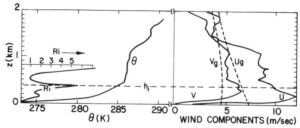

Fig. 6.8 Observed wind and potential temperature profiles under very stable (sporadic turbulence) conditions at night during the Wangara Experiment. [From Deardorff (1978).]

mum in Ri. The Ri profile indicates that continuous turbulence may be expected only in the lower half of the PBL, where Ri < 0.25.

The sporadic turbulence regime is more common in the NBL over land. The typical observed profiles of U, V, Θ, and Ri in the very stable regime are shown in Fig. 6.8. Note that both the wind components attain maximum values at low levels. The PBL depth based on the maximum in the

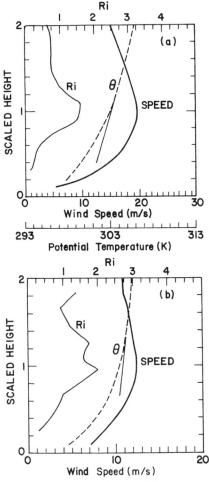

Fig. 6.9 Composite profiles of wind speed, potential temperature, and Richardson number scaled with respect to the low-level jet height at (a) O'Neill, Nebraska, during the Great Plains Experiment and (b) Hay, Australia, during the Wangara Experiment. [After Mahrt *et al.* Copyright © (1979) by D. Reidel Publishing Company. Reprinted by permission.]

wind speed profile is only about half of the depth of the surface inversion layer. The Ri profile indicates that the layer of possible continuous turbulence in which Ri < 0.25 is very shallow indeed and in the bulk of the NBL turbulence is likely to be intermittent, patchy, and weak.

The occurrence of a low-level jet is a common phenomenon in the NBL (see, e.g., Fig. 6.9). The jet can become highly intensified when thermal or slope winds oppose the surface geostrophic wind. Such intensified low-level jets have frequently been observed over the Great Plains region of the United States. Figure 6.9 shows low-level jets in the composite wind speed profiles obtained from observations during the Wangara Experiment and the Great Plains Field Program (Lettau and Davidson, 1957).

Other characteristic features of the stably stratified PBL are large directional shear (wind veering) and small PBL depth, both of which are manifestations of the effects of stability. The veering of wind with height is better illustrated through a representation of wind components in the form of wind hodograph, which is the locus of the end points of velocity vectors at various heights. Figure 6.10 presents such hodographs from a 32-m tower at Plateau Station, Antarctica. These are actually averages of many observed wind profiles grouped under eight stability classes (from the least stable class 1 to the most stable class 8) and three periods or seasons. The corresponding temperature profiles have already been given in Fig. 5.5. Note that with increasing stability, wind veering with height also increases. For extremely stable classes changes in wind direction with

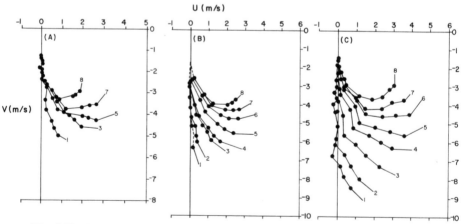

Fig. 6.10 Average observed wind hodographs at Plateau Station, Antarctica, for the three periods grouped under different stability classes. (A) Sunlight periods, (B) transitional periods, (C) dark periods. The components shown are in the so-called geotriptic coordinate system. Stability classes: 1, 2, 3, 4, 5, 6, 7, and 8; heights: 0.5, 1, 2, 4, 8, 12, 16, 20, 24, and 32 m. [After Lettau *et al.* (1977).]

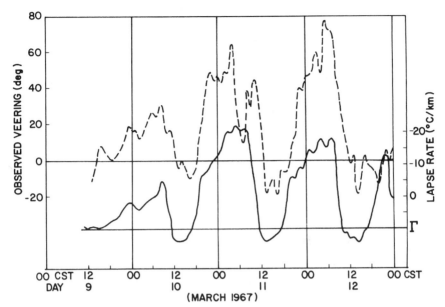

Fig. 6.11 Observed relationship between wind veering (---) and lapse rate (——) near Oklahoma City, Oklahoma. [After Gray and Mendenhall (1973).]

height can be detected even at low levels of 1 or 2 m, so that the surface layer becomes very shallow under these conditions.

The close relationship between wind veering and stability or temperature lapse rate is further demonstrated by observations given in Fig. 6.11. Note that wind veering increases with increasing stability (negative lapse rate); the negative values (backing of wind) under superadiabatic conditions are probably due to the influence of thermal winds.

6.6 DIURNAL VARIATIONS

Diurnal variations of wind speed and wind direction in the PBL have been inferred from observations collected at different locations. The evaluation of wind profiles in the course of a diurnal cycle may show large day-to-day variations due to changes in the synoptic weather situation and the surface energy balance. When averaged over long periods of time (order of a month or longer), however, the diurnal variations are better discerned.

Figure 6.12 shows the diurnal variations of mean wind speed (averaged over a period of 1 yr) observed from an instrumented 500-m tower near

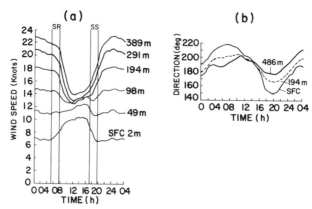

Fig. 6.12 Diurnal variations of (a) mean wind speed and (b) mean wind direction, on an annual basis, for various height levels in the PBL near Oklahoma City, Oklahoma. [After Crawford and Hudson (1973).]

Oklahoma City, Oklahoma. The ranges of times of sunrise and sunset, as well as the observation heights, are indicated in the figure. Note that the near-surface wind speed increases sharply after sunrise, attains a broad maximum in early afternoon, and decreases sharply near sunset. The diurnal wave at the higher level in the surface layer is similar, but with a reduced amplitude. The increase in the strength of surface winds following the morning inversion breakup is due to more rapid and efficient transfer of momentum from aloft through the evolving unstable or convective PBL in the daytime.

Above the surface layer, the diurnal wave becomes nearly 180° out of phase and wind speed decreases sharply following sunrise, attains a minimum value around noon, and increases in the afternoon and evening hours. The amplitude of the wave increases with height and attains its maximum value somewhere in the middle of the convective PBL. Although these observations did not extend to higher levels, the amplitude of diurnal variation of wind speed is expected to decrease farther up and vanish at the maximum height of the daytime PBL. Similar diurnal patterns of wind speed at various heights in the PBL are shown in Fig. 6.13, which is based on 40-day averages of pibal observed winds during the Wangara expedition at a smooth rural site in Australia.

It can be inferred from the diurnal evolution of potential temperature and specific humidity profiles shown in Chapter 5 and, to some extent, from the wind profiles also, that PBL depth has a strong diurnal variation, especially during the daytime under clear skies. The unstable or convective PBL is usually capped by an inversion and the height of the inversion base z_i is considered to be a measure of the PBL depth h (actually, h is

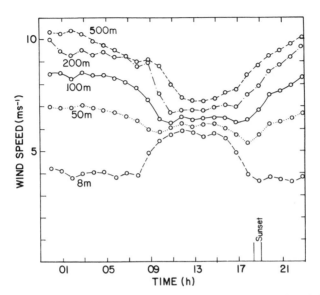

Fig. 6.13 Diurnal variations of 40-day-averaged wind speeds at various heights in the PBL during the Wangara Experiment. [After Mahrt (1981).]

5–30% larger than z_i, if one includes the interfacial transition layer in the former). The diurnal evolution of the height of the lowest inversion base on a typical day during the 1973 Minnesota experiment is shown in Fig. 6.14. Also represented in the same figure is the diurnal variation of the surface heat flux H_0 whose cumulative input (integration with respect to time) to the PBL is primarily responsible for the growth of z_i or h.

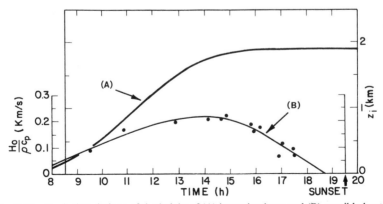

Fig. 6.14 Typical variations of the height of (A) inversion base and (B) sensible heat flux during a daytime period in the Minnesota Experiment. [After Kaimal *et al.* (1976).]

6.7 APPLICATIONS

The knowledge of wind distribution in the PBL has the following practical applications:

- Determining the rate of dissipation of the kinetic energy in the lower atmosphere
- Determining local and regional transports of pollutants and air trajectories in the lower atmosphere
- Estimating momentum flux or shear stress through the PBL and the surface drag
- Estimating wind energy potential and designing wind power-generating systems
- Designing tall buildings, towers, and bridges for wind loads
- Designing wind shelters and other protective measures

PROBLEMS AND EXERCISES

1. Derive the thermal wind equations from the geostrophic wind relations, the hydrostatic equation, and the equation of state for the lower atmosphere, and indicate when and why the terms involving the vertical temperature gradient may be neglected.
2. Show that the net horizontal friction force per unit volume of a fluid element in the PBL is $\partial\tau/\partial z$ and that it need not be opposite to the velocity vector $\mathbf{V}$. How is the friction force related to the ageostrophic wind vector?
3. The following measurements of mean winds and temperatures were made from the 200-m mast at Cabauw in The Netherlands on a September night:

Height (m)	Wind speed (m sec^{-1})	Wind direction (deg)	Temperature (°C)
5	2.2	194	8.58
10	2.7	194	8.81
20	3.5	202	9.08
40	5.4	217	10.15
80	8.2	234	12.19
120	9.1	240	12.96
160	8.7	246	12.86
200	7.8	250	12.63

Geostrophic wind speed at the surface = 5.94 m sec^{-1}
Geostrophic wind direction at the surface = 263°
Horizontal temperature gradient toward east = -2.69×10^{-5} K m^{-1}
Horizontal temperature gradient toward north = 8.85×10^{-6} K m^{-1}
Boundary layer depth = 100 m
Surface pressure = 1000 mbar

(a) Plot on a graph the wind speed, wind direction, and potential temperature profiles.
(b) Plot the wind hodograph, using the geographical coordinate system, and indicate geostrophic wind vectors at the surface and the top of the PBL. Compare the actual wind with the geostrophic wind at 100 m.
(c) Calculate and plot the wind component profiles, taking the x axis along the near-surface (5 m) wind.

4. (a) Using the wind component and potential temperature profiles of the previous problem, determine the magnitudes of wind shear, potential temperature gradient, and Richardson number at the heights of 7.5, 15, 30, 60, 100, 140, and 180 m.
(b) Plot Ri as a function of height and indicate the probable top of the PBL based on the critical Richardson number of 0.5.

5. The following observations are for a midlatitude ($f = 10^{-4}$ sec^{-1}) PBL during the afternoon convective conditions:

Height[a] (m)	Wind speed[b] (m sec^{-1})	Wind direction[c] (deg)
10	9.0	250
100	15.5	253
1000	17.0	258
1500	15.0	260

[a] The PBL height = 1500 m.
[b] The surface geostrophic wind speed = 20.0 m sec^{-1}.
[c] The surface geostrophic wind direction = 275°.

(a) Calculate the average magnitudes of the geostrophic wind shear and the actual wind shear in the PBL, assuming geostrophic balance at the top of the PBL.
(b) Compute the total wind veering and frictional veering across the PBL, as well as the cross-isobar angle of the near-surface winds.

(c) Compute the pressure gradient, Coriolis, and friction forces per unit mass ($\rho = 1.2 \text{ kg m}^{-3}$) at the 10- and 1000-m levels and vectorially represent (to scale) the force balance at each level.

6. Calculate the magnitudes and directions of pressure gradient, Coriolis, and friction forces per unit mass of a fluid element at the top of a nocturnal surface layer where the actual winds are 7.07 m sec^{-1} from the west, the geostrophic wind speed is 10 m sec^{-1}, and the cross-isobar angle is 45° (take $f = 10^{-4}$ sec^{-1}).

Chapter 7 | An Introduction to Viscous Flows

7.1 INVISCID AND VISCOUS FLOWS

In the preceding two chapters, our treatment of temperature, humidity, and wind distributions in the PBL was largely empirical and based on rather limited observations. A better understanding of vertical profiles of wind, temperature, and humidity and their relationships to the important exchanges of momentum, heat, and moisture through the PBL has been gained through applications of mathematical, statistical, and semiempirical theories. A brief description of some of the fundamentals of viscous flows and turbulence and the associated theory will be given in this and the following chapters.

For theoretical treatments, fluid flows are commonly divided into two broad categories, namely, inviscid and viscous flows. In an inviscid or ideal fluid the effects of viscosity are completely ignored, i.e., the fluid is assumed to have no viscosity, and the flow is considered to be nonturbulent. Inviscid flows are smooth and orderly, and the adjacent fluid layers can easily slip past each other or against solid surfaces without any friction or drag. Consequently, there is no mixing and no transfer of momentum, heat, and mass across the moving layers. Such properties can only be transported along the streamlines through advection. The inviscid flow theory obviously results in some serious dilemmas and inconsistencies when applied anywhere close to solid surfaces or density interfaces. Such dilemmas can be resolved only by recognizing the presence of boundary layers or interfacial mixing layers in which the effects of viscosity or turbulence cannot be ignored. Far away from the boundaries and density interfaces, however, the fluid viscosity can be ignored and the inviscid or ideal flow model provides a very good approximation to many real fluid flows encountered in geophysical and engineering applications. Extensive applications of this are given in books on hydrodynamics and geophysical fluid dynamics (see, e.g., Lamb, 1932; Pedlosky, 1979).

VISCOSITY AND ITS EFFECTS

Fluid viscosity is a molecular property which is a measure of the internal resistance of the fluid to deformation. All real fluids, whether liquids or gases, have finite viscosities associated with them. An important manifestation of the effect of viscosity is that fluid particles adhere to a solid surface as they come in contact with the latter and consequently there is no relative motion between the fluid and the solid surface. If the surface is at rest, the fluid motion right at the surface must also vanish. This is called

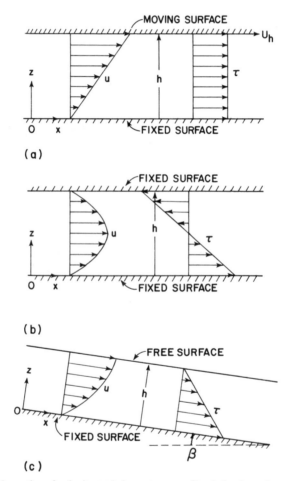

Fig. 7.1 Schematics of velocity and shear stress profiles in laminar plane-parallel flows: (a) Couette flow, (b) channel flow, (c) free surface gravity flow.

the no-slip boundary condition, which is also applicable at the interface of the two fluids with widely different densities (e.g., air and water).

Within the fluid flow, viscosity is responsible for the frictional resistance between adjacent fluid layers. The resistance force per unit area is called the shearing stress, because it is associated with the shearing motion (variation of velocity) between the layers. A simple demonstration of this is provided by the smooth, streamlined, laminar flow between two large parallel planes, one fixed and the other moving at a slow constant speed U_h, which are separated by a small distance h (see Fig. 7.1a). At sufficiently small values of U_h and h, laminar flow can be maintained and the velocity within the fluid varies linearly from zero at the fixed plane to U_h at the moving plane, so that the velocity gradient $\partial u/\partial z = U_h/h$, everywhere in the flow. From observations in such a flow, Newton found the relationship that shearing stress is proportional to the rate of strain or velocity gradient (fluids following a linear relationship between the stress and the rate of strain are known as Newtonian fluids), that is,

$$\tau = \mu(\partial u/\partial z) \qquad (7.1)$$

Where the coefficient of proportionality μ is called the dynamic viscosity of the fluid. In fluid flow problems, it is more convenient to use the kinematic viscosity $\nu \equiv \mu/\rho$, which has dimensions of $L^2 T^{-1}$.

In particular, for the gaseous fluids, such as air, a more rigorous theoretical derivation of the relationship between the shearing stress and velocity gradient can be obtained from the kinetic theory (see, e.g., Sutton, 1953), which clearly shows viscosity to be a molecular property which depends on the temperature and pressure in the fluid (this dependence is, however, weak and is usually ignored in most atmospheric applications).

Equation (7.1) is strictly valid only for a unidirectional flow. In most real flows, in general, spatial variations of velocity in different directions give rise to shear stresses in different directions. More general constitutive relations between the components of shearing stress and velocity gradient at any point in the fluid are

$$\tau_{xy} = \tau_{yx} = \mu(\partial u/\partial y + \partial v/\partial x)$$

$$\tau_{xz} = \tau_{zx} = \mu(\partial u/\partial z + \partial w/\partial x) \qquad (7.2)$$

$$\tau_{yz} = \tau_{zy} = \mu(\partial v/\partial z + \partial w/\partial y)$$

in which the first member of the subscript to τ denotes the direction normal to the plane of the shearing stress and the second member denotes the direction of the stress. Equation (7.2) implies that in Newtonian fluids, shear stresses are proportional to the applied strain rates or deformations (represented by quantities in parentheses). Here, stresses and deformations are all instantaneous quantities at a point in the flow.

Another important effect of viscosity is the dissipation of kinetic energy of the fluid motion, which is constantly converted into heat. Therefore, in order to maintain the motion, the energy has to be continuously supplied externally, or converted from potential energy, which exists in the form of pressure and density gradients in the fluid.

Although, the above-mentioned effects of viscosity are felt, in varying degrees, in all real fluid flows, at all times they are found to be particularly significant only in certain regions and in certain types of flows. Such flows are called viscous flows, as opposed to the inviscid flows mentioned earlier. Examples of such flows are boundary layers, mixing layers, jets, plumes, and wakes, which are dealt with in many books on fluid mechanics (see, e.g., Batchelor, 1970; Townsend, 1976).

7.2 LAMINAR AND TURBULENT FLOWS

All viscous flows can broadly be classified as laminar and turbulent flows, although an intermediate category of transition between the two has also been recognized. A laminar flow is characterized by smooth, orderly, and slow motion in which adjacent layers (*laminae*) of fluid slide past each other with very little mixing and transfer (only at the molecular scale) of properties across the layers. The main difference between the laminar flows and the inviscid flows introduced earlier is the importance of viscous and other molecular transfers of momentum, heat, and mass in the former. In laminar flows, the flow field and the associated temperature and concentration fields are regular and predictable and vary only gradually in space and time.

In sharp contrast to laminar and inviscid flows, turbulent flows are highly irregular, almost random, three-dimensional, highly rotational, dissipative, and very diffusive (mixing) motions. In these, all the flow and scalar properties exhibit highly irregular variations (fluctuations) in both time and space, with a wide range of temporal and spatial scales. For example, turbulent fluctuations of velocity in the atmospheric boundary layer typically occur over time scales ranging from 10^{-3} to 10^4 sec and the corresponding spatial scales from 10^{-3} to 10^4 m—more than a millionfold range of scales. Due to their nearly random nature (there is some order, persistence, and correlation between fluctuations in time and space), turbulent motions cannot be predicted or calculated exactly as functions of time and space; one usually deals with their average statistical properties.

Although there are many examples and applications of laminar flows in industry, in the laboratory, and in biological systems, their occurrence in natural environments, particularly the atmosphere, is rare and confined to the so-called viscous sublayers over smooth surfaces (e.g., ice, mud flats, relatively undisturbed water, and tree leaves). Most fluid flows encountered in nature and engineering applications are turbulent. In particular, the various small-scale motions in the lower atmosphere are turbulent. The turbulent mixing layer may vary in depth from a few tens of meters in clear, calm, nocturnal cooling conditions to several kilometers in highly disturbed (stormy) weather conditions involving deep penetrative convection. In the upper troposphere and stratosphere also, extensive regions of clear air turbulence are found to occur. Some meteorologists consider all atmospheric motions right up to the scale of general circulation as turbulent. But, then, one has to distinguish between the small-scale three-dimensional turbulence of the type we encounter in micrometeorology, and the large-scale "two-dimensional turbulence." There are fundamental differences in some of their properties and mechanisms of energy transfers up or down the scales. We will only be concerned with the former here.

Turbulent, too, are the flows in upper oceans (e.g., oceanic mixed layer), lakes, rivers, and channels; practically all flows in liquid and gas pipelines; in boundary layers over moving aircraft, missiles, ships, and boats; in wakes of structures, propeller blades, turbine blades, projectiles, bullets, and rockets; in jets of exhaust gases and liquids; and in chimney and cooling tower plumes. Thus turbulence is literally all around us, particularly in the form of the atmospheric boundary layer, from which we can escape only briefly. The next and the following chapters will be devoted to fundamentals of turbulence and their application to micrometeorology. But first we discuss the basic equations for viscous motion, which are valid for both laminar and turbulent flows.

7.3 EQUATIONS OF MOTION

Mathematical treatments of fluid flows, including those in the atmosphere, are almost always based on the equations of motion, which are mathematical expressions of the fundamental laws of the conservation of mass, momentum, and energy. For example, considerations of mass conservation in an elemental volume of fluid leads to the equation of continuity, which for an incompressible fluid, in a Cartesian coordinate system, is given by

$$\partial u/\partial x + \partial v/\partial y + \partial w/\partial z = 0 \tag{7.3}$$

The continuity equation imposes an important constraint on the fluid motion such that the divergence of velocity must be zero at all times and at every point in the flow. The assumption of incompressible fluid or flow is quite well justified in micrometeorological applications.

The application of Newton's second law of motion, or consideration of momentum conservation in an elemental volume of fluid, leads to the so-called Navier–Stokes equations, which, in a Cartesian frame of reference tied to the surface of the rotating earth with x and y axes in the horizontal and z axis in the vertical, are

$$\frac{Du}{Dt} - fv = -\frac{1}{\rho}\frac{\partial p}{\partial x} + \nu\nabla^2 u$$

$$\frac{Dv}{Dt} + fu = -\frac{1}{\rho}\frac{\partial p}{\partial y} + \nu\nabla^2 v \qquad (7.4)$$

$$\frac{Dw}{Dt} + g = -\frac{1}{\rho}\frac{\partial p}{\partial z} + \nu\nabla^2 w$$

Where the total derivative and Laplacian operator are defined as

$$D/Dt \equiv \partial/\partial t + u(\partial/\partial x) + v(\partial/\partial y) + w(\partial/\partial z) \qquad (7.5)$$

$$\nabla^2 \equiv \partial^2/\partial x^2 + \partial^2/\partial y^2 + \partial^2/\partial z^2 \qquad (7.6)$$

The derivation of the above equations of motion is outside the scope of this text and can be found elsewhere (Haltiner and Martin, 1957; Dutton, 1976). The physical interpretation and significance of each term can be given as follows. Terms in Eq. (7.4) represent accelerations or forces per unit mass of the fluid element in x, y, and z direction, respectively. On the left-hand sides, the first terms are called the inertia terms, because they represent the inertia forces arising due to local and advective accelerations on a fluid element. The second terms in the equations of horizontal motion represent the Coriolis accelerations or forces which apparently act on the fluid element due to the earth's rotation. The rotational term is insignificant in the equation of vertical motion and has been omitted from the same. Instead, the gravitational acceleration term appears in the equation for vertical motion. On the right-hand sides of Eq. (7.4), the first terms represent the pressure gradient forces and the second terms are viscous or friction forces on the fluid element.

When fluid viscosity or friction terms can be ignored, the equations of motion reduce to the so-called Euler's equations for an inviscid or ideal fluid. The solutions of these equations for a variety of density or temperature stratifications and initial and boundary conditions now form the rich body of literature in classical hydrodynamics (Lamb, 1932) and geophysical fluid dynamics (Pedlosky, 1979). Euler's equations of motion are considered to be adequate for describing the behavior of atmosphere and oceans outside of any boundary, interfacial, and mixing layers in which the flows are likely to be turbulent.

For a mathematical description of viscous flows, one has to use the complete set of the Navier–Stokes equations, which are nonlinear partial differential equations of second order and are extremely difficult (often impossible) to solve. The combination of nonlinear inertia terms and the viscous terms is responsible for the extreme difficulty and, in many cases, the intractability of solutions. Analytical solutions have been found only for limited cases of very low-speed laminar or creeping flows when nonlinear terms are absent or can be simplified. Some of these "exact" solutions are given in the following sections in order to introduce the reader to certain classical fluid flows.

7.4   PLANE-PARALLEL FLOWS

The simplest viscous flows are the steady, one-dimensional, laminar flows between two infinite parallel planes. Since velocity varies only normal to the planes and there is no motion across the planes, the inertia terms in the Navier–Stokes equations are zero. For small-scale laboratory flows, the earth's rotational effects (Coriolis terms) can also be ignored. Furthermore, choosing the x axis in the direction of flow and the z axis normal to the plane boundaries, the equations of motion reduce to

$$d^2u/dz^2 = (1/\mu)(\partial p/\partial x) \qquad (7.7)$$

which can easily be solved for different flow situations (boundary conditions).

7.4.1   PLANE-COUETTE FLOW

Consider the laminar flow between two parallel boundaries, one fixed and the other moving in the x direction at a constant velocity of U_h, separated by a small distance h, as depicted in Fig. 7.1a. The relevant no-slip boundary conditions are

$$u = 0, \qquad \text{at} \quad z = 0$$
$$u = U_h, \qquad \text{at} \quad z = h \qquad (7.8)$$

The solution of Eq. (7.7) satisfying the above boundary conditions is given by

$$u = U_h \frac{z}{h} - \frac{1}{2\mu} \frac{\partial p}{\partial x} z(h - z) \qquad (7.9a)$$

or, in the dimensionless form,

$$\frac{u}{U_h} = \frac{z}{h} - \frac{h^2}{2\mu U_h} \frac{\partial p}{\partial x} \frac{z}{h} \left(1 - \frac{z}{h}\right) \qquad (7.9b)$$

Note that the velocity profile in this so-called plane-Couette flow is a combination of linear and parabolic profiles. For the special case of zero-pressure gradient ($\partial p/\partial x = 0$), the velocity profile becomes linear, i.e.,

$$u/U_h = z/h \qquad (7.10)$$

and the shear stress is equal to the surface shear stress $\tau_0 = \mu U_h/h$ everywhere in the flow. This flow is entirely forced by the moving surface, or the relative motion between the two parallel surfaces, and is depicted in Fig. 7.1a.

7.4.2   PLANE-POISEUILLE OR CHANNEL FLOW

When both of the parallel bounding surfaces are fixed, the appropriate boundary conditions for the unidirectional flow between them are

$$u = 0, \quad \text{at} \quad z = 0$$
$$u = 0, \quad \text{at} \quad z = h \qquad (7.11)$$

Then, the solution of Eq. (7.7) yields a parabolic velocity profile

$$u = -\frac{1}{2\mu} \frac{\partial p}{\partial x} z(h - z) \qquad (7.12)$$

which implies a linear shear stress profile

$$\tau = \mu \frac{\partial u}{\partial z} = -\frac{1}{2} \frac{\partial p}{\partial x} (h - 2z) \qquad (7.13)$$

with the maximum value at the surface $\tau_0 = -(h/2)(\partial p/\partial x)$. Note that in order to have a steady flow through the channel, there must be a constant negative pressure gradient in the direction of the flow. The velocity and shear stress profiles in a plane channel flow are schematically shown in Fig. 7.1b.

The solution of simplified equations of motion in a cylindrical coordinate system yields similar velocity and stress profiles in a circular pipe (Hagen–Poiseuille) flow.

7.4.3 GRAVITY FLOW DOWN AN INCLINED PLANE

Here we consider a unidirectional liquid flow with a free surface or other gravity flows over uniformly sloping surfaces. The relevant equation of motion with the x axis in the direction of flow (down the slope β) and the z axis normal to the surface (Fig. 7.1c) is

$$d^2u/dz^2 = -(g/\nu) \sin \beta \tag{7.14}$$

in which we have included the gravity force term but have ignored the pressure gradient. The boundary conditions are

$$
\begin{aligned}
u = 0, &\quad \text{at} \quad z = 0 \\
du/dz = 0, &\quad \text{at} \quad z = h
\end{aligned} \tag{7.15}
$$

The upper boundary condition implies zero shear stress at the free surface, or at $z = h$, where h represents the constant thickness of the frictional shear layer.

The solution of Eq. (7.14) satisfying the above boundary conditions is given by

$$u = \frac{g \sin \beta}{2\nu} z(2h - z) \tag{7.16}$$

which is again a parabolic profile with the maximum velocity at $z = h$

$$u_h = \frac{gh^2}{2\nu} \sin \beta \tag{7.17}$$

The shear stress profile is linear with the maximum value at the surface

$$\tau_0 = \rho g h \sin \beta \tag{7.18}$$

These profiles are schematically shown in Fig. 7.1c. Note that both the maximum velocity and the surface shear stress are proportional to the sine of the slope angle.

7.5 EKMAN LAYERS

In this section, we consider the laminar Ekman boundary layers on rotating surfaces, particularly those of the atmosphere and oceans, aside from the conditions of their existence.

7.5.1 EKMAN LAYER BELOW THE SEA SURFACE

First, we examine the problem of drift currents set up at or just below the sea surface, in response to a steady wind stress forcing. For the sake of simplicity, surface waves are being ignored and so are the horizontal pressure and density gradients in water. Then, the equations of horizontal motion [Eq. (7.4)] reduce to

$$-fv = \nu \frac{d^2u}{dz^2}; \qquad fu = \nu \frac{d^2v}{dz^2} \tag{7.19}$$

Assuming the x axis is in the direction of the applied surface stress τ_0 and the z axis is pointed vertically upward, the boundary conditions to be satisfied are

$$\mu \frac{du}{dz} = \tau_0, \qquad \mu \frac{\partial v}{\partial z} = 0, \qquad \text{at} \quad z = 0$$
$$u \to 0, \qquad v \to 0, \qquad \text{as} \quad z \to -\infty \tag{7.20}$$

The two equations can be combined in a single equation for the complex current $c = u + iv$, by multiplying the second part of Eq. (7.19) by $i = \sqrt{-1}$ and adding it to the first. The resulting equation

$$d^2c/dz^2 - i(f/\nu)c = 0 \tag{7.21}$$

has the solution satisfying the boundary condition at infinity

$$c = u + iv = A \exp[a(1 + i)z] \tag{7.22}$$

where $a = (f/2\nu)^{1/2}$ and A is a complex constant determined from the surface boundary condition to be

$$A = (\tau_0/2a\mu)(1 - i) \tag{7.23}$$

Substituting from Eq. (7.23) into Eq. (7.22) and separating into real and imaginary parts, one obtains

$$u = (\tau_0/\sqrt{2}a\mu)e^{az} \cos(az - \pi/4)$$
$$v = (\tau_0/\sqrt{2}a\mu)e^{az} \sin(az - \pi/4) \tag{7.24}$$

Note that according to the above solution, the current has a maximum speed of $\tau_0/\sqrt{2}a\mu$ at the surface and is directed 45° in a clockwise sense from the direction of applied stress. With increasing depth (negative z) the current speed decreases exponentially and the current direction rotates in a clockwise sense. A common method of representing the current speed and direction as a function of depth below the surface is to plot a velocity hodograph as in Fig. 7.2a. Note that the velocity hodograph represented by Eq. (7.24) is a spiral; it is known as the Ekman spiral, after the Swedish oceanographer V. W. Ekman who first derived the above solution.

Although, the induced current theoretically disappears only at an infinite depth, in practice, the influence of surface stress forcing becomes insignificant at a depth of the order of $a^{-1} = (2\nu/f)^{1/2}$. Conventionally, the

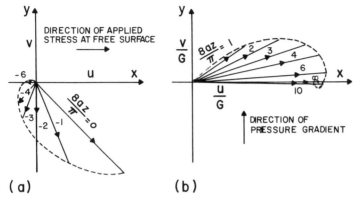

Fig. 7.2 Velocity hodographs in laminar Ekman layers: (a) below a free surface where tangential stress is applied, (b) above a rigid surface where a constant pressure gradient is applied. [From Batchelor (1970).]

Ekman layer depth h_E is defined as the depth where the current direction becomes exactly opposite to the surface current direction. According to Eq. (7.24), this happens at $z = -\pi a^{-1}$, so that $h_E = \pi(2\nu/f)^{1/2}$; at this depth the magnitude of the current has fallen to a small fraction ($e^{-\pi} \simeq 0.04$) of its surface value.

Another parameter of interest is the net volume flux of water across vertical planes due to induced currents in the Ekman layer, which is given by the integral

$$\int_0^{-\infty} (u + iv) \, dz = -i(\tau_0/\rho f) \tag{7.25}$$

Thus, the net flux is normal to the direction of applied stress and is proportional to the stress, but independent of fluid viscosity.

7.5.2 EKMAN LAYER AT A RIGID SURFACE

Another example of Ekman layer is that due to a uniform pressure gradient in the atmosphere near the surface or in the ocean near the bottom boundary. For the sake of simplicity, again, the surface is assumed to be flat and uniform and the flow is considered laminar. In the absence of inertia terms, the equations of motion to be solved are

$$-fv = -\frac{1}{\rho}\frac{\partial p}{\partial x} + \nu\frac{d^2u}{dz^2}$$

$$fu = -\frac{1}{\rho}\frac{\partial p}{\partial y} + \nu\frac{d^2v}{dz^2} \tag{7.26}$$

Expressing pressure gradients in terms of geostrophic velocity components using Eq. (6.1), Eq. (7.26) can be written as

$$-f(v - V_g) = \nu(d^2/dz^2)(u - U_g)$$
$$f(u - U_g) = \nu(d^2/dz^2)(v - V_g)$$

(7.27)

in which U_g and V_g have been taken as height-independent parameters. The appropriate boundary conditions, assuming geostrophic balance outside the Ekman layer, are

$$u = 0, \qquad v = 0, \qquad \text{at} \quad z = 0$$
$$u \to U_g, \qquad v \to V_g, \qquad \text{as} \quad z \to \infty$$

(7.28)

The previous set of equations [Eq. (7.27)] is similar to Eq. (7.19), and the solution, following the earlier solution, is

$$u - U_g = -e^{-az}[U_g \cos(az) + V_g \sin(az)]$$
$$v - V_g = e^{-az}[U_g \sin(az) - V_g \cos(az)]$$

(7.29)

This is a general solution which is independent of the orientation of the horizontal coordinate axes. A more specific and simpler solution is usually given by taking the x axis to be oriented with the geostrophic wind vector, so that $U_g = G$ and $V_g = 0$ and Eq. (7.29) can be written as

$$u = G[1 - e^{-az} \cos(az)]$$
$$v = Ge^{-az} \sin(az)$$

(7.30)

The normalized wind hodograph (Ekman spiral), according to Eq. (7.30), is shown in Fig. 7.2b. It shows that, in the Northern Hemisphere, the wind vector rotates in a clockwise sense as the height increases, and that the angle between the surface wind or stress and the geostrophic wind is 45° (this is also the cross-isobar angle of the flow near the surface), irrespective of stability. The Ekman layer depth again turns out to be $h_E = \pi(2\nu/f)^{1/2}$, which in middle latitudes ($f \simeq 10^{-4}$ sec^{-1}) is only about 1.7 m (the depth of the wind-induced oceanic Ekman spiral is about 0.5 m).

There is no observational evidence for the occurrence of such shallow laminar Ekman layers in the atmosphere or oceans. Therefore, the above solutions would be of academic interest only, unless ν is replaced by a much higher effective (eddy) viscosity K to match the theoretical Ekman layer depth with that observed under a given set of conditions. Even then, certain features of the above solution remain inconsistent with observations. For example, the cross-isobar angle of the atmospheric flow near the surface is highly variable, as discussed in Chapter 6, and the theoretical value of $\alpha_0 = 45°$ is found more as an exception rather than the rule.

Another inconsistent feature of the Ekman solution is almost linear velocity profile near the surface, while observations indicate approximately
logarithmic or log-linear velocity profiles in the surface layer. Slightly
modified solutions with an appropriate choice of effective viscosity and
lower boundary conditions (in the form of specified wind speed or wind
direction) at the top of the surface layer have been given in the literature
and largely remove the above-mentioned limitations and inconsistencies
of the original Ekman solution and show a better correspondence with
observed wind profiles.

7.6 DEVELOPING BOUNDARY LAYERS

In the simple cases of plane-parallel flows discussed in Sections 7.4 and
7.5, inertial accelerations or forces are zero and there is a balance between the pressure gradient and friction forces, or between pressure gradient, Coriolis, and friction forces. As a result, thickness of the affected
fluid layer does not change in the flow direction or x–y plane. In developing viscous flows, such as boundary layers, wakes, and jets, the thickness
of the layer in which viscous or friction effects are important changes in
the direction of flow, and inertial forces are as important, if not more, as
the viscous forces. The ratio of the two defines the Reynolds number
$Re = UL/\nu$, where U and L are the characteristic velocity and length
scales (there may be more than one length scale characterizing the flow).
This ratio is a very important characteristic of any viscous flow and indicates the relative importance of inertial forces as compared to viscous
forces.

Atmospheric boundary layers developing over most natural surfaces
are characterized by very large (10^6–10^9) Reynolds numbers. Boundary
layers encountered in engineering practice also have fairly large Reynolds
numbers ($Re = 10^3$–10^6, based on the boundary layer thickness as the
length scale and the ambient velocity just outside the boundary layer as
the velocity scale). One would expect that in such large Reynolds number
flows the nonlinear inertia terms in the equations of motion will be far
greater in magnitude than the viscous terms, at least in a gross sense. Still,
it turns out that the viscous effects cannot be ignored in order to satisfy
the no-slip boundary condition and to provide a smooth transition,
through the boundary layer, from zero velocity at the surface to finite
ambient velocity outside the boundary layer. Recognizing this, Prandtl
(1905) first proposed an important boundary layer hypothesis which states
that, under rather broad conditions, viscosity effects are significant in
layers adjoining solid boundaries and in certain other layers (e.g., mixing

layers and jets), the thicknesses of which approach zero as the Reynolds number of the flow approaches infinity, and are small outside these layers. This hypothesis has been applied to a variety of flow fields and is supported by many observations. The thickness of a boundary or mixing layer should be looked at in relation to the distance over which it develops. This explains the existence of relatively thick boundary layers in the atmosphere, in spite of very large Reynolds numbers characterizing the same.

The boundary layer hypothesis, for the first time, explained most of the dilemmas of inviscid flow theory, and clearly defined regions of the flow where such a theory may not be applicable and other regions where it would provide close simulations of real fluid flows. It also provided a practically useful definition of the boundary layer as the layer in which the fluid velocity makes a transition from that of the boundary (zero velocity, in the case of a fixed boundary) to that appropriate for an ambient (inviscid) flow. In theory, as well as in practice, the approach to ambient velocity is often very smooth and asymptotic, so that some arbitrariness or ambiguity is always involved in defining the boundary layer thickness. In engineering practice, the outer edge of the boundary layer is usually taken where the mean velocity has attained 99% of its ambient value. This definition is found to be quite unsatisfactory in meteorological applications, where other more useful definitions of the boundary layer thickness have been used (e.g., the Ekman layer thickness defined earlier).

Prandtl's boundary layer hypothesis is found to be valid for laminar as well as turbulent boundary layers. The fact that the boundary layer is thin compared with the distance over which it develops along a boundary allows for certain approximations to be made in the equations of motion. These boundary layer approximations, also due to Prandtl, amount to the following simplifications for the viscous diffusion terms in Eq. (7.4):

$$\nu\nabla^2 u \simeq \nu(\partial^2 u/\partial z^2)$$

$$\nu\nabla^2 v \simeq \nu(\partial^2 v/\partial z^2) \qquad (7.31)$$

$$\nu\nabla^2 w \simeq \nu(\partial^2 w/\partial z^2)$$

which follow from the neglect of velocity gradients parallel to the boundary in comparison to those normal to it.

THE FLAT-PLATE LAMINAR BOUNDARY LAYER

Let us consider the simple case of a steady, two-dimensional boundary layer developing over a thin flat plate placed in an otherwise steady, uniform stream of fluid, with streamlines of the ambient flow parallel to

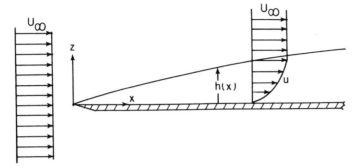

Fig. 7.3 Schematic of a developing boundary layer over a flat plate.

the plate (see Fig. 7.3). This is an important classical case of fluid flow which provides a standard for comparison with boundary layers developing on slender bodies, such as aircraft wings, ship hulls, and turbine blades. Except in a small region near the edge of the plate, where boundary layer approximations may not be valid, the relevant boundary layer equations to be solved are

$$u(\partial u/\partial x) + w(\partial u/\partial z) = \nu(\partial^2 u/\partial z^2)$$

$$\partial u/\partial x + \partial w/\partial z = 0$$
(7.31)

with the boundary conditions

$$u = w = 0, \quad \text{at} \quad z = 0$$

$$u \to U_\infty, \quad \text{as} \quad z \to \infty$$
(7.32)

$$u = U_\infty, \quad \text{at} \quad x = 0, \quad \text{for all } z$$

where U_∞ is the uniform ambient velocity just outside the boundary layer.

Some idea about the growth of the boundary layer with distance downstream of the leading edge can be obtained by requiring that the inertial and viscous terms of Eq. (7.31) be of the same order of magnitude in the boundary layer, i.e.,

$$u(\partial u/\partial x) \sim \nu(\partial^2 u/\partial z^2)$$

or

$$U_\infty^2/x \sim \nu U_\infty/h^2$$

or

$$h \sim (\nu x/U_\infty)^{1/2} = x/\mathrm{Re}_x^{1/2}$$
(7.33)

where $\mathrm{Re}_x = U_\infty x/\nu$ is a local Reynolds number. The above result indicates that the boundary layer thickness grows in proportion to the square root of the distance and that the ratio h/x decreases inversely proportional to the square root of the local Reynolds number. This supports the idea of a thin boundary layer in a large Reynolds number flow.

Equation (7.31) can be solved numerically. The resulting velocity profiles at various distances from the edge of the plate turn out to be similar in the sense that they collapse onto a single curve if the normalized longitudinal velocity u/U_∞ is plotted as a function of the normalized distance z/h, or $z/(\nu x/U_\infty)^{1/2}$, from the surface. If, to begin with, one assumes a similarity solution of the form

$$u/U_\infty = f'(\eta) = \partial f/\partial \eta \qquad (7.34)$$

where $\eta = z(U_\infty/\nu x)^{1/2}$, it is easy to show that the partial differential equations [Eq. (7.31)] yield a single ordinary differential equation

$$f''' + \tfrac{1}{2}ff'' = 0 \qquad (7.35)$$

where prime denotes differentiation with respect to η. The boundary conditions [Eq. (7.32)] can be transformed to

$$f = f' = 0, \qquad \text{at} \quad \eta = 0$$
$$f' \rightarrow 1, \qquad\qquad \text{as} \quad \eta \rightarrow \infty \qquad (7.36)$$

The normalized velocity profile obtained from the numerical solution of the above equations is shown in Fig. 7.4. Note that the profile near the

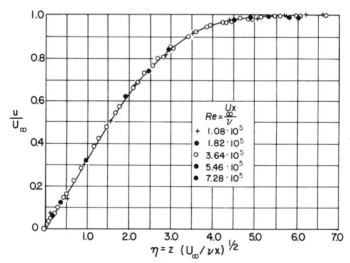

Fig. 7.4 Comparison of theoretical and observed velocity profiles in the laminar flat-plate boundary layer. [From Schlichting (1960).]

plate surface is nearly linear, similar to the velocity profiles in Ekman layers and plane-parallel flows (this linearity of profiles seems to be a common feature of all laminar flows). The shear stress on the plate surface is given by

$$\tau_0 = \mu(\partial u/\partial z)_{z=0} = 0.33\rho U_\infty^2 \mathrm{Re}_x^{-1/2} \qquad (7.37)$$

according to which the friction or drag coefficient, $C_\mathrm{D} \equiv \tau_0/\rho U_\infty^2$, varies inversely proportional to the square root of the local Reynolds number. The boundary layer thickness (defined as the value of z where $u = 0.99 U_\infty$) is given by

$$h \simeq 5(\nu x/U_\infty)^{1/2} \qquad (7.38)$$

Many measurements in flat-plate laminar boundary layers have confirmed these theoretical results (see, e.g., Schlichting, 1960).

7.7   HEAT TRANSFER IN FLUIDS

In Section 4.4 we derived the one-dimensional equation of heat conduction in a solid medium and pointed out how it can be generalized to three dimensions. Heat transfer in a still fluid is no different from that in a solid medium, the molecular conduction or diffusion being the only mechanism of transfer. In a moving fluid, however, heat is more efficiently transferred by fluid motions while thermal conduction is a relatively slow process.

7.7.1   FORCED CONVECTION

Consideration of conservation of energy in an elementary fluid volume, in a Cartesian coordinate system, leads to the following equation of heat transfer in a low-speed, incompressible flow in which temperature or heat is considered only a passive admixture not affecting in any way the dynamics.

$$DT/Dt = \alpha_h \nabla^2 T \qquad (7.39)$$

where α_h is the molecular thermal diffusivity of the fluid and the total derivative has the usual meaning as defined by Eq. (7.5) and incorporates the important contribution of fluid motions.

Equation (7.39) describes the variation of temperature for a given velocity field and can be solved for the appropriate boundary conditions and flow situations. Analytical solutions are possible only for certain types of laminar flows. For two-dimensional thermal boundary layers, jets, and plumes the boundary layer approximations are used to simplify Eq. (7.39) in the form

$$u(\partial T/\partial x) + w(\partial T/\partial z) = \alpha_h(\partial^2 T/\partial z^2) \tag{7.40}$$

which is similar to the simplified boundary layer equation of motion.

Analogous to the Reynolds number, one can define the Peclet number $Pe = UL/\alpha_h$, which represents the ratio of the advection terms to the molecular diffusion terms in the energy equation. Instead of Pe, it is often more convenient to use the Prandtl number $Pr = \nu/\alpha_h$, which is only a property of the fluid and not of the flow; note that

$$Pe = Re \cdot Pr \tag{7.41}$$

For air and other diatomic gases, $Pr \simeq 0.7$ and is nearly independent of temperature, so that the Peclet number is of the same order of magnitude as the Reynolds number.

The difference in the temperature at a point in the flow and at the boundary is usually normalized by the temperature difference ΔT across the entire thermal layer and is represented as a function of normalized coordinates of the point, the appropriate Reynolds number, and the Prandtl number. The heat flux on a boundary is usually represented by the dimensionless ratio, called the Nusselt number

$$Nu = H_0 L/\alpha_h \Delta T \tag{7.42}$$

or, the heat transfer coefficient

$$C_H = H_0/\rho c_p U \Delta T \tag{7.43}$$

which is also called the Stanton number. Note that the above parameters are related (through their definitions) as

$$C_H = Nu/RePr = Nu/Pe \tag{7.44}$$

Many calculations, as well as observations of heat transfer in pipes, channels, and boundary layers have been used to establish empirical correlations between the Nusselt number or the heat transfer coefficient as a function of the Reynolds and Prandtl numbers. Some of these have found practical applications in micrometeorological problems, particularly those dealing with heat and mass transfer to or from individual leaves (see, e.g., Monteith, 1973; Lowry, 1970; Gates, 1980).

7.7.2 FREE CONVECTION

In free or natural convection flows, temperature cannot be considered as a passive admixture, because inhomogeneities of the temperature field in the presence of gravity give rise to significant buoyant accelerations that affect the dynamics of the flow, particularly in the vertical direction. This is usually the case in the atmosphere. The equations of motion and thermodynamic energy are slightly modified to allow for the buoyancy effects of temperature or density stratification. It is generally assumed that there is a reference state of the atmosphere at rest characterized by temperature T_0, density ρ_0, and pressure p_0, which satisfy the hydrostatic equation

$$\partial p_0 / \partial z = -\rho_0 g \qquad (7.45)$$

and that, in the actual atmosphere, the deviations in these properties from their reference values ($T_1 = T - T_0$, $\rho_1 = \rho - \rho_0$ and $p_1 = p - p_0$) are small as compared to the values for the reference atmosphere ($T_1 \ll T_0$, $\rho_1 \ll \rho_0$, etc). These Boussinesq assumptions lead to the approximation

$$(1/\rho)(\partial p/\partial z) + g \simeq (1/\rho_0)(\partial p_1/\partial z) + (g/\rho_0)\rho_1 \qquad (7.46)$$

which can be used in the vertical equation of motion. In an ideal gas, such as air, changes in density are simply related to changes in temperature as

$$\rho_1 = -\beta \rho_0 T_1 \simeq -(\rho_0/T_0)T_1 \qquad (7.47)$$

in which the coefficient of thermal expansion $\beta = -(1/\rho_0)(\partial \rho/\partial T)_p \simeq 1/T_0$. After substituting from Eqs. (7.46) and (7.47) in the equation of vertical motion [Eq. (7.4)], we have

$$Dw/Dt = -(1/\rho_0)(\partial p_1/\partial z) + (g/T_0)T_1 + \nu \nabla^2 w \qquad (7.48)$$

Here, the second term on the right-hand side is the buoyant acceleration due to the deviation of temperature from the reference state. The equations of horizontal motion remain unchanged; alternatively, one can replace the pressure gradient terms $(1/\rho)(\partial p/\partial x)$ and $(1/\rho)(\partial p/\partial y)$ by $(1/\rho_0)(\partial p_1/\partial x)$ and $(1/\rho_0)(\partial p_1/\partial y)$, respectively. A more appropriate form of the thermodynamic energy equation, for micrometeorological applications, is

$$D\theta/Dt = \alpha_h \nabla^2 \theta \qquad (7.49)$$

in which θ is the potential temperature.

The above equations are valid for both the stable and unstable stratifications. In the particular case of true free convection over a flat plate (horizontal or vertical) in which the motions are entirely generated by surface heating, the relevant dimensionless parameters on which temperature and velocity fields depend are the Prandtl number and the Grashof number

$$\text{Gr} = (g/T_0)(L^3 \Delta T/\nu^2) \qquad (7.50)$$

Instead of Gr, one may also use the Rayleigh number

$$\text{Ra} = (g/T_0)(L^3 \Delta T/\nu \alpha_h) = \text{Gr} \cdot \text{Pr} \qquad (7.51)$$

Both the Grashof and Rayleigh numbers are expected to be large in the atmosphere, although true free convection in the sense of no geostrophic wind forcing may not be a frequent occurrence.

7.8 APPLICATIONS

Since, micrometeorology deals primarily with the phenomena and processes occurring within the atmospheric boundary layer, some familiarity with the fundamentals of viscous fluid flows is essential. In particular, the basic differences between viscous and inviscid flows and between laminar and turbulent viscous flows should be recognized. The Navier–Stokes equations of motion and the thermodynamic energy equation and difficulties of their solution must be familiar to the students of micrometeorology. Introduction to some of the fundamentals of fluid flow and heat transfer may provide a useful link between dynamic meteorology and fluid mechanics. A direct application of this would be in micrometeorological studies of momentum, heat, and mass transfer between the atmosphere and the physical and biological elements (e.g., snow, ice, and water surfaces and plant leaves, animals, and organisms).

PROBLEMS AND EXERCISES

1. In what regions or situations in the atmosphere may the inviscid flow theory not be applicable and why?
2. What are the basic difficulties in the solution of the Navier–Stokes equations of motion for laminar viscous flows?
3. Using the simplified equation of motion in a cylindrical coordinate system

$$(1/r)(d/dr)[r(du/dr)] = (1/\mu)(\partial p/\partial x)$$

   obtain expressions for velocity distribution and shear stress in a laminar pipe flow.
4. In the gravity flow down an inclined plane, determine the volume flow rate Q across the plane per unit width of the plane and express the layer depth h as a function of Q and the slope angle β.
5. For the atmospheric Ekman layer, obtain the solutions of Eq. (7.27) in a coordinate system with the x axis along the surface wind and the lower boundary conditions $u = u_s$, $v = 0$, at $z = h_s$.
6. Plot the velocity hodograph and u and v profiles obtained in Problem 5 above as functions of z from 10 to 1000 m for the following conditions:

Geostrophic wind speed, $G = 10$ m sec^{-1}
Surface wind speed, $u_s = 7$ m sec^{-1} at $z = h_s = 10$ m
Effective viscosity, $\nu = K = 2$ m^2 sec^{-1}
Coriolis parameter, $f = 10^{-4}$ sec^{-1}.

What is the Ekman layer thickness and the cross-isobar angle of the surface flow?

7. (a) By direct substitution verify that Eq. (7.29) is a solution to Eq. (7.27) which satisfies the boundary conditions [Eq. (7.28)].
   (b) Using the above solution [Eq. (7.29)], obtain the corresponding expressions for the horizontal shear stress components τ_{zx} and τ_{zy}.
   (c) Show that in a coordinate system with the x axis parallel to the surface shear stress (this implies that $\tau_{zx} = \tau_0$ and $\tau_{zy} = 0$, at $z = 0$), $U_g = G/\sqrt{2}$, $V_g = -G/\sqrt{2}$, and $\tau_0 = \rho G(\nu f)^{1/2}$.
   (d) In the same coordinate system write down the expressions for the normalized velocity components (u/G and v/G) and the normalized shear stress components (τ_{zx}/τ_0 and τ_{zy}/τ_0).
   (e) Using the above expressions, calculate and plot the vertical profiles of the normalized velocity and shear stress components as functions of az from 0 to 2π.
   (f) What conclusions can you draw from the above profiles? Comment on their applicability to the real atmosphere, if ν can be replaced by an effective viscosity K.

8. Starting from Eqs. (7.31) and (7.34), derive Eq. (7.35) for the dimensionless velocity profile in a flat-plate boundary layer.

9. Derive Eq. (7.46) for the stratified atmosphere, using the Boussinesq assumption that the deviations of thermodynamic variables from their reference state values are small.

Chapter 8 | Fundamentals of Turbulence

8.1 INSTABILITY OF FLOW AND TRANSITION TO TURBULENCE

8.1.1 TYPES OF INSTABILITIES

In Chapter 5 we introduced the concept of static (gravitational) stability or instability and showed how the parameter $s = (g/T_{v0})(\partial\Theta_v/\partial z) \simeq -(g/\rho_0)(\partial\rho/\partial z)$ may be used as a measure of static stability of an atmospheric (fluid) layer. This was based on the simple criterion whether vertical motions of fluid parcels were suppressed or enhanced by the buoyancy force arising from density differences between the parcel and the environment. In particular, when $s < 0$, the fluid layer is gravitationally unstable; the parcel moves farther and farther away from its equilibrium position.

Another type of flow instability is the dynamic or hydrodynamic instability. A flow is considered dynamically stable if perturbations introduced in the flow, either intentionally or inadvertently, are found to decay with time or distance in the direction of flow, and eventually get suppressed altogether. It is dynamically unstable if the perturbations grow in time or space and irreversibly alter the nature of the basic flow. A flow may be stable with respect to very small (infinitesimal) perturbations but might become unstable if large (finite) amplitude disturbances are introduced in the flow.

A simple example of dynamically unstable flow is that of two inviscid fluid layers moving initially at different velocities so that there is a sharp discontinuity at the interface. The development of the so-called Kelvin–Helmholtz instability can be demonstrated by assuming a slight perturbation (e.g., in the form of a standing wave) of the interface and the resulting perturbations in pressure (in accordance with the Bernoulli equation $u^2/2 + p/\rho = $ constant) at various points across the interface. The forces induced by pressure perturbations tend to progressively increase the amplitude of the wave, no matter how small it was to begin with, thus irreversibly modifying the interface and the flow in its vicinity. In real fluids, of course, the processes leading to the development of mixing layers are more complicated, but the initiation of the Kelvin–Helmholtz instability at the interface is qualitatively similar to that in an inviscid fluid.

Laminar flows in channels, tubes, and boundary layers, introduced in Chapter 7, are found to be dynamically stable (i.e., they can be maintained as laminar flows) only under certain restrictive conditions (e.g., Reynolds number less than some critical value Re_c and a relatively disturbance-free environment), which can be realized only in carefully controlled laboratory facilities. For example, by carefully eliminating all extraneous disturbances, laminar flow in a smooth tube has been realized up to $Re \simeq 10^5$, while under ordinary circumstances the critical Reynolds number (based on mean velocity and tube radius) is only about 1000. Similarly, the critical Reynolds numbers for the stability (maintenance) of laminar channel and boundary layer flows are also of the same order of magnitude. For other flows, such as jets and wakes, instabilities occur at much lower Reynolds numbers. Thus, all laminar flows are expected to become dynamically unstable and, hence, cannot be maintained as such, at sufficiently large Reynolds numbers ($Re > Re_c$) and in the presence of disturbances ordinarily present in the environment.

The above Reynolds number criterion for dynamic stability does not mean that all inviscid flows are inherently unstable, because zero viscosity implies an infinite Reynolds number. Actually, many nonstratified (uniform density) inviscid flows are found to be stable with respect to small perturbations, unless their velocity profiles have discontinuities or inflexion points. The development of Kelvin–Helmoltz instability along the plane of discontinuity in the velocity profile has already been discussed. Even without a sharp discontinuity, the existence of an inflexion point in the velocity profile constitutes a necessary condition for the occurrence of instability in a neutrally stratified flow; it is also a sufficient condition for the amplification of disturbances. This inviscid instability mechanism also operates in viscous flows having points of inflexion in their velocity profiles. It is for this reason that laminar jets, wakes, mixing layers, and separated boundary layers become unstable at low Reynolds numbers. The laminar Ekman layers are also inherently unstable because their velocity profiles have inflexion points. In all such flows, both the viscous and inviscid instability mechanisms operate simultaneously and, consequently, the stability criteria are more complex.

Gravitational instability is another mechanism through which both the inviscid and viscous streamlined flows can become turbulent. Flows with fluid density increasing with height are dynamically unstable, particularly at large Reynolds numbers (Re > Re_c) or Rayleigh numbers (Ra > Ra_c). Stably stratified flows with weak negative density gradients and/or strong velocity gradients can also become dynamically unstable, if the Richardson number is less than its critical value Ri_c = 0.25 (this value is based on a number of theoretical as well as experimental investigations). The Richardson number criterion is found to be extremely useful for identifying possible regions of active or incipient turbulence in stably stratified atmospheric and oceanic flows. If the local Richardson number in a certain region is less than 0.25 and there is also an inflexion point in the velocity profile (e.g., in jet streams and mixing layers), the flow there is likely to be turbulent. This criterion is found to be generally valid in stratified boundary layers, as well as in free shear flows, and is often used in forecasts of clear-air turbulence in the upper troposphere and lower stratosphere for aircraft operations.

8.1.2   METHODS OF INVESTIGATION

The stability of a physically realizable laminar flow can be investigated experimentally and the approximate stability criteria can be determined empirically. This approach has been used by experimental fluid dynamists ever since the classical experiments in smooth glass tubes carried out by Osborne Reynolds. Systematic and carefully controlled experiments are required for this purpose (see Monin and Yaglom, 1971, Chap. 1). Empirical observations to test the validity of the Richardson number criterion have also been made by meteorologists and oceanographers.

Another, more frequently used approach is theoretical, in which small-amplitude wavelike perturbations are superimposed on the original laminar or inviscid flow and the conditions of stability or instability are examined through analytical or numerical solutions of perturbation equations. The analysis is straightforward and simple, so long as the amplitudes of perturbations are small enough for the nonlinear terms in the equations to be dropped out or linearized. It becomes extremely complicated and even intractable in the case of finite-amplitude perturbations. Thus, most theoretical investigations give only some weak criteria of the stability of flow with respect to small-amplitude (in theory, infinitesimal) perturbations. Investigations of flow stability with respect to finite amplitude perturbations are being attempted now using new approaches to the numerical solution of nonlinear perturbation equations on large computers. Comprehensive reviews of the literature on stability analyses of viscous and inviscid flows are given by Monin and Yaglom (1971) and Pedlosky (1979).

The importance of stability analysis arises from the fact that any solution to the equations of motion, whether exact or approximate, cannot be considered physically realizable unless it can be shown that the flow represented by the solution is also stable with respect to small perturbations. If the basic flow is found to be unstable in a certain range of parameters, it cannot be realized under those conditions, because naturally occurring infinitesimal disturbances will cause instabilities to develop and alter the flow field irreversibly.

8.1.3 TRANSITION TO TURBULENCE

Experiments in controlled laboratory environments have shown that several distinctive stages are involved in the transition from laminar flow (e.g., the flat-plate boundary layer) to turbulence. The initial stage is the development of primary instability which, in simple cases, may be two dimensional. The primary instability produces secondary motions which are generally three dimensional and become unstable themselves. The subsequent stages are the amplification of three-dimensional waves, the development of intense shear layers, and the generation of high-frequency fluctuations. Finally, "turbulent spots" appear more or less randomly in space and time, grow rapidly, and merge with each other to form a field of well-developed turbulence. The above stages are easy to identify in a flat-plate boundary layer, because changes occur as a function of increasing Reynolds number with distance from the leading edge. In two-dimensional channel and pipe flows, transition to turbulence occurs more suddenly and explosively over the whole length of the channel or pipe as the Reynolds number is increased beyond its critical value.

Mathematically, the details of transition from initially laminar or inviscid flow to turbulence are rather poorly understood. Much of the theory is linearized, valid for small disturbances, and cannot be used beyond the initial stages. Even the most advanced nonlinear theories dealing with finite amplitude disturbances cannot handle the later stages of transition, such as the development of "turbulent spots." A rigorous mathematical treatment of transition from turbulent to laminar flow is also lacking, of course, due to the lack of a generally valid and rigorous theory of turbulence.

Some of the turbulent flows encountered in nature and technology may not go through a transition of the type described above and are produced as such (turbulent). Flows in ordinary pipes, channels, and rivers, as well as in atmospheric and oceanic boundary layers, are some of the commonly occurring examples of such flows. Other cases, such as clear air turbulence in the upper atmosphere and patches of turbulence in the stratified ocean, are obviously the results of instability and transition processes. Transitions from laminar flow to turbulence and vice versa continuously occur in the upper parts of the nocturnal boundary layer, throughout the night. Therefore, micrometeorologists have a deep and abiding interest in understanding these transition processes.

8.2 THE GENERATION AND MAINTENANCE OF TURBULENCE

It is not easy to identify the origin of turbulence. In an initially nonturbulent flow, the onset of turbulence may occur suddenly through a breakdown of streamlined flow in certain localized regions (turbulent spots). The cause of the breakdown, as pointed out in the previous section, is an instability mechanism acting upon the naturally occurring disturbances in the flow. Since turbulence is transported downstream in the manner of any other fluid property, repeated breakdowns are required to maintain a continuous supply of turbulence, and instability is an essential part of this process.

Once turbulence is generated and becomes fully developed in the sense that its statistical properties achieve a steady state, the instability mechanism is no longer required (although it may still be operating) to sustain the flow. This is particularly true in the case of shear flows, where shear provides an efficient mechanism of converting mean flow energy to turbulent kinetic energy (TKE). Similarly, in unstably stratified flows, buoyancy provides a mechanism of converting potential energy of stratification into turbulent kinetic energy (the reverse occurs in stably stratified flows). The two mechanisms of turbulence generation become more evident from the so-called shear production (S) and buoyancy production (B) terms in the TKE equation

$$d(\text{TKE})/dt = S + B - D + T_r \qquad (8.1)$$

the derivation of which is outside the scope of this text. The TKE equation also shows that there is a continuous dissipation (D) of energy by viscosity in any turbulent flow and there may be transport (T_r) of energy from or to other regions of flow. Thus, in order to maintain turbulence, one or more of the generating mechanisms must be active continuously. For example, in the daytime atmospheric boundary layer, turbulence is produced both by shear and buoyancy (the relative contribution of each depends on many factors). Shear is the only effective mechanism of producing turbulence in the nocturnal boundary layer, in low-level jets, and in regions of upper troposphere and stratosphere where negative buoyancy actually suppresses turbulence. Near the earth's surface wind shears become particularly intense and effective, because the velocity must vanish at the surface and air flow has to go around the various surface inhomogeneities. For this reason, shear-generated turbulence is always present in the atmospheric surface layer.

Richardson (1920) originally proposed a particularly simple condition for the maintenance of turbulence in stably stratified flows, viz., that the rate of production of turbulence by shear must be equal to or greater than the rate of destruction by buoyancy. That is, the Richardson number, which turns out to be the ratio of the buoyancy to shear production, must be smaller than or equal to unity. An important assumption implied in deriving this condition was that there is no viscous dissipation of energy on reaching the critical condition for the sudden decay of turbulence. On the other hand, subsequent observations of the transition from turbulent to laminar flow have indicated viscous dissipation to remain significant and a much lower value of the critical Richardson number (Ri_c), in the range of 0.2 to 0.5 [the values from 0.20 to 0.30 are more frequently used in the literature, although it has not been possible to determine Ri_c very precisely (see Arya, 1972)]. There is no requirement that this should be the same as the critical Richardson number for the transition from laminar to turbulent flow, which has been determined more rigorously and precisely from theoretical stability analyses.

8.3 GENERAL CHARACTERISTICS OF TURBULENCE

Even though turbulence is a rather familiar notion, it is not easy to define precisely. In simplistic terms, we refer to very irregular and chaotic motions as turbulent. But some wave motions (e.g., on an open sea surface) can be very irregular and nearly chaotic, but they are not turbulent. Perhaps it would be more appropriate to mention some general characteristics of turbulence.

1. Irregularity or randomness. This makes any turbulent motion essentially unpredictable. No matter how carefully the conditions of an experiment are reproduced, each realization of the flow is different and cannot be predicted in detail. The same is true of the numerical simulations (based on Navier–Stokes equations), which are found to be highly sensitive to even minute changes in initial and boundary conditions. For this reason, a statistical description of turbulence is invariably used in practice.

2. Three-dimensionality and rotationality. The velocity field in any tur-

bulent flow is three-dimensional (we are excluding here the so-called two-dimensional or geostrophic turbulence, which includes all large-scale atmospheric motions) and highly variable in time and space. Consequently, the vorticity field is also three-dimensional and flow is highly rotational.

3. Diffusivity or ability to mix properties. This is probably the most important property, so far as applications are concerned. It is responsible for the efficient diffusion of momentum, heat, and mass (e.g., water vapor, CO_2, and various pollutants) in turbulent flows. Macroscale diffusivity of turbulence is usually many orders of magnitude larger than the molecular diffusivity. The former is a property of the flow while the latter is a property of the fluid. Turbulent diffusivity is largely responsible for the evaporation in the atmosphere, as well as for the spread (dispersion) of pollutants released in the atmospheric boundary layer. It is also responsible for the increased frictional resistance of fluids in pipes and channels, around aircraft and ships, and on the earth's surface.

4. Dissipativeness. The kinetic energy of turbulent motion is continuously dissipated (converted into internal energy or heat) by viscosity. Therefore, in order to maintain turbulent motion, the energy has to be supplied continuously. If no energy is supplied, turbulence decays rapidly.

5. Multiplicity of scales of motion. All turbulent flows are characterized by a wide range (depending on the Reynolds number) of scales or eddies. The transfer of energy from the mean flow into turbulence occurs at the upper end of scales (large eddies), while the viscous dissipation of turbulent energy occurs at the lower end (small eddies). Consequently, there is a continuous transfer of energy from the largest to the smallest scales. Actually, it trickles down through the whole spectrum of scales or eddies in the form of a cascade process. The energy transfer processes in turbulent flows are highly nonlinear and are not well understood.

Of the above characteristics, rotationality, diffusivity, and dissipativeness are the properties which distinguish a three-dimensional random wave motion from turbulence. The wave motion is nearly irrotational, nondiffusive, and nondissipative.

8.4 MEAN AND FLUCTUATING VARIABLES

In a turbulent flow velocity, temperature and other variables vary irregularly in time and space; it is therefore common practice to consider these variables as sums of mean and fluctuating parts, e.g.,

$$\tilde{u} = U + u$$
$$\tilde{v} = V + v$$
$$\tilde{w} = W + w \qquad (8.2)$$
$$\tilde{\theta} = \Theta + \theta$$

in which the left-hand side represents an instantaneous variable (denoted by a tilde) and the right-hand side its mean (denoted by a capital letter) and fluctuating (denoted by a lower case letter) parts. The decomposition of an instantaneous variable in terms of its mean and fluctuating parts is called Reynolds decomposition.

There are several types of means or averages used in theory and practice. The most commonly used in the analysis of observations from fixed instruments is the time mean, which, for a continuous record (time series), can be defined as

$$F = \frac{1}{T} \int_0^T \tilde{f}(t) \, dt$$

where $\tilde{f}$ is any variable or a function of variables, F is the corresponding time mean, and T is the length of record or sampling time over which averaging is desired. In certain flows, such as the atmospheric boundary layer, the choice of an optimum sampling time or the averaging period T is not always clear. It should be sufficiently long to ensure a stable average and to incorporate the effects of all the significantly contributing large eddies in the flow. On the other hand, T should not be too long to mask what may be considered as real trends (e.g., the diurnal variations) in the flow. In the analysis of micrometeorological observations, the optimum averaging time may range between 10^3 and 10^4 sec, depending on the height of observation, the PBL height, and stability.

Another type of averaging used, particularly in the analyses of aircraft, radar, and sodar observations, is the spatial average along the observation path. The optimum length, area, or volume over which the spatial averaging is performed depends on the spatial scales of significant large eddies in the flow.

Finally, the type of averaging which is almost always used in theory, but rarely in practice, is the ensemble or probability mean. It is an arithmatical average of a very large (approaching to infinity) number of realizations of a variable or a function of variables, which are obtained by repeating the experiment over and over again under the same general conditions. It is quite obvious that the ensemble averages would be nearly impossible to obtain under the varying weather conditions in the atmosphere, over which we have little or no control. Even in a controlled laboratory environment, it would be very time consuming to repeat an experiment many times to obtain ensemble averages.

The fact that different types of averages are used in theory and experiments requires that we should know about the conditions in which time or space averages might become equivalent to ensemble averages. It has been found that the necessary and sufficient conditions for the time and ensemble means to be equal are that the process (in our case, flow) be stationary (i.e., the averages be independent of time) and the averaging period be very large ($T \rightarrow \infty$) (for a more detailed discussion of these, see Monin and Yaglom, 1971). It should suffice to say that these conditions cannot be strictly satisfied in the atmosphere so that the above-mentioned equivalence of averages can only be approximate. The corresponding conditions for the equivalence of spatial and ensemble averages are spatial homogeneity (independence of averages to spatial coordinates in one or more directions) and very large averaging paths or areas. These conditions are even more restrictive and difficult to satisfy in micrometeorological applications.

Irrespective of the type of averaging used, it is obvious from Reynolds decomposition that a fluctuation is the deviation of an instantaneous variable from its mean. By definition, then, the mean of any fluctuating variable is zero ($\bar{u} = 0$, $\bar{v} = 0$, etc.), so that there are compensating negative fluctuations for positive fluctuations on the average. Here, an overbar over a variable also denotes its mean.

The relative magnitudes of mean and fluctuating parts of variables in the atmospheric boundary layer depend on the type of variable, observation height relative to the PBL height, atmospheric stability, the type of surface, and other factors. In the surface layer, during undisturbed weather conditions, the magnitudes of vertical fluctuations, on the average, are much larger than the mean vertical velocity, the magnitudes of horizontal velocity fluctuations are of the same order or less than the mean horizontal velocity, and the magnitudes of the fluctuations in thermodynamic variables are at least two orders of magnitude smaller than their mean values.

8.5 VARIANCES AND TURBULENT FLUXES

The distribution of mean variables, such as mean velocity components, temperature, etc., can tell much about the mean structure of a flow, but little or nothing about the turbulent exchanges taking place in the flow. Various statistical measures are used to study and represent the turbulence structure. These are all based on statistical analyses of turbulent fluctuations observed in the flow.

The simplest measures of fluctuation levels are the variances $\overline{u^2}, \overline{v^2}, \overline{w^2}$ [these three are often combined to define the turbulent kinetic energy per unit mass $\frac{1}{2}(\overline{u^2} + \overline{v^2} + \overline{w^2})$], $\overline{\theta^2}$, etc., and standard deviations $\sigma_u = (\overline{u^2})^{1/2}$, etc. The ratios of standard deviations of velocity fluctuations to the mean wind speed are called turbulence intensities (e.g., $i_u = \sigma_u/U$, $i_v = \sigma_v/U$ and $i_w = \sigma_w/U$), which are measures of relative fluctuation levels in different directions (velocity components). Observations indicate that turbulence intensities are typically less than 10% in the nocturnal boundary layer, 10–20% in a near-neutral surface layer, and greater than 20% in unstable and convective boundary layers. Turbulence intensities are generally largest near the surface and decrease with height in stable and near-neutral boundary layers. Under convective conditions, however, the secondary maxima in turbulence intensities often occur at the inversion base and sometimes in the middle of the CBL.

Even more important in turbulent flows are the covariances $\overline{uw}, \overline{vw}, \overline{\theta w}$, etc., which are directly related to and sometimes referred to as turbulent fluxes of momentum, heat, etc. As covariances, they are averages of products of two fluctuating variables and depend on the correlations between the variables involved. These can be positive, negative, or zero, depending on the type of flow and symmetry conditions. For example, if u and w are the velocity fluctuations in the x and z directions, respectively, their product uw will also be a fluctuating quantity but with a nonzero mean; $\overline{uw}$ will be positive if u and w are positively correlated, and negative if the two variables are negatively correlated. A better measure of the correlation between two variables is provided by the correlation coefficient

$$r_{uw} = \overline{uw}/(\overline{u^2}\,\overline{w^2})^{1/2} = \overline{uw}/\sigma_u\sigma_w \tag{8.3}$$

whose value always lies between -1 and 1, according to Schwartz's inequality $|\overline{uw}| \leq \sigma_u\sigma_w$. Similarly, one can define correlation coefficients for all the covariances to get an idea of how well correlated different variables are in a turbulent flow. There is still some order to be found in an apparently chaotic motion which is responsible for all the important exchange processes taking place in the flow.

In order to see a clear connection between covariances and turbulent fluxes, let us consider the vertical flux of a scalar with a variable concentration (mass per unit volume) $\bar{c}$ in the flow. The flux of any scalar in a given direction is defined as the amount of the scalar per unit time per unit area normal to that direction. It is quite obvious that the velocity component in the direction of flux is responsible for the transport and hence for the flux. For example, it is easy to see that in the vertical direction the flux at any instant is $\tilde{c}\tilde{w}$ and the mean flux with which we are normally concerned is $\overline{\tilde{c}\tilde{w}}$. Further writing $\tilde{c} = C + c$ and $\tilde{w} = W + w$ and following the Reynolds averaging rules, it can be shown that

$$\overline{\tilde{c}\tilde{w}} = \overline{(C + c)(W + w)} = CW + \overline{cw} \tag{8.4}$$

Thus, the total scalar flux can be represented as a sum of the mean transport (transport by mean motion) and the turbulent transport. The latter, also called the turbulent flux, is usually the dominant transport term and always of considerable importance in a turbulent flow. If the scalar under consideration is the potential temperature $\bar{\theta}$, or enthalpy $\rho c_p \bar{\theta}$, the corresponding turbulent flux in the vertical is $\overline{\theta w}$, or $\rho c_p \overline{\theta w}$. In this way, the covariance $\overline{\theta w}$ may be interpreted as the turbulent heat flux (in kinematic units). Similarly, $\overline{uw}$ may be considered as the vertical turbulent flux of horizontal (x direction) momentum or, equivalently, the horizontal (x direction) flux of vertical momentum, as $\overline{uw} = \overline{wu}$.

According to Newton's second law of motion the rate of change (flux) of momentum is equal to force per unit area or stress, so that one can also express turbulent stresses as

$$\tau_{xz} = -\rho\overline{uw}, \qquad \tau_{yz} = -\rho\overline{vw}, \quad \text{etc.} \tag{8.5}$$

These are also called Reynolds stresses, which in most turbulent flows are found to have much larger magnitudes than the corresponding mean viscous stresses $\mu(\partial U/\partial z + \partial W/\partial x)$, $\mu(\partial V/\partial z + \partial W/\partial y)$, etc. The latter can usually be ignored, except within the extremely thin viscous sublayers.

8.6 EDDIES AND SCALES OF MOTION

It is a common practice to speak in terms of eddies when turbulence is described qualitatively. An eddy is by no means a clearly defined structure or feature of the flow which can be isolated and followed through, in order to study its behavior. It is rather an abstract concept used mainly for qualitative descriptions of turbulence. An eddy may be considered akin to a vortex or a whirl in common terminology. Turbulent flows are highly rotational and have all kinds of vortexlike structures (eddies) buried in them. However, eddies are not simple two-dimensional circulatory motions of the type in an isolated vortex, but are believed to be complex, three-dimensional structures. Any analogy between turbulent eddies and vortices can only be very rough and qualitative.

On the basis of flow visualization studies, statistical analyses of turbulence data, and some of the well-accepted theoretical ideas, it is believed that a turbulent flow consists of a hierarchy of eddies of a wide range of sizes (length scales), from the smallest that can survive the dissipative action of viscosity to the largest that is allowed by the flow geometry. The range of eddy sizes increases with the Reynolds number of the overall mean flow. In particular, for the ABL, the typical range of eddy sizes is 10^{-3} to 10^3 m.

Of the wide and continuous range of scales in a turbulent flow, a few have a special significance and are used to characterize the flow itself. The one is the characteristic large-eddy scale or macroscale of turbulence, which represents the length (l) or time scale of eddies receiving the most energy from the mean flow. The other is the characteristic small-eddy scale or microscale of turbulence, which represents the length (η) or time scale of most dissipating eddies. The ratio l/η is found to be proportional to $Re^{3/4}$ and is typically 10^5 for the atmospheric PBL.

The macroscale, or simply the scale of turbulence, is generally comparable (in order of magnitude) to the characteristic scale of the mean flow, such as the boundary layer thickness or channel depth, and does not depend on the molecular properties of the fluid. On the other hand, the microscale depends on the fluid viscosity ν, as well as on the rate of energy dissipation ε. Dimensional considerations further lead to the defining relationship

$$\eta \equiv \nu^{3/4} \varepsilon^{-1/4} \tag{8.6}$$

In stationary or steady-state conditions, the rate at which the energy is dissipated is exactly equal to the rate at which the energy is supplied from mean flow into turbulence. This leads to an inviscid estimate for the energy dissipation $\varepsilon \sim u^3/l$, where u is the characteristic velocity scale of turbulence (large eddies), which can be defined in terms of the turbulent kinetic energy

$$\text{TKE} \equiv \tfrac{1}{2}(\overline{u^2} + \overline{v^2} + \overline{w^2}) = \tfrac{3}{2}u^2 \tag{8.7}$$

Some investigators define the turbulence velocity scale such that $\tfrac{1}{2}u^2$ represents the turbulent kinetic energy (e.g., Tennekes and Lumley, 1972).

It has been recognized for a long time that in large Reynolds number flows almost all of the energy is supplied to large eddies, while almost all of it is eventually dissipated by small eddies. The transfer of energy from large (energy-containing) to small (energy-dissipating) eddies occurs through a cascade-type process involving the whole range of intermediate size eddies. This idea was originally suggested by Lewis Richardson in 1922 in the form of a parody

Big whorls have little whorls,
Which feed on their velocity;
And little whorls have lesser whorls,
And so on to viscosity.

This may be considered to be the qualitative picture of turbulence structure in a nutshell and the concept of energy transfer down the scale. The actual mechanism and quantitative aspects of energy transfer are very complicated and are outside the scope of this text. Smaller eddies are assumed to be created through an instability and breakdown of larger eddies and the energy transfer from larger to smaller eddies presumably occurs during the breakdown process. Since small eddies are created after many successive breakdowns of larger and larger eddies, it can be argued that the former are very far removed from the initial process of the generation of macroscale turbulence by the mean flow and, hence, can have no preferred orientation with respect to the direction of mean shear or buoyancy. Thus, the small-scale structure may be expected to have some universal attributes common to all turbulent flows. This idea of universal similarity and isotropy of small-scale turbulence was first expounded by A. N. Kolmogorov in 1941 (see Monin and Yaglom, 1971) and has become the cornerstone of theory and observations of turbulence.

8.7 APPLICATIONS

A knowledge of the fundamentals of turbulence is essential to any qualitative or quantitative understanding of the turbulent exchange processes occurring in the PBL as well as in the free atmosphere. The statistical descriptions in terms of variances and covariances of turbulent fluctuations, as well as in terms of eddies and scales of motion, provide the bases for further quantitative analyses of observations. At the same time the basic concepts of the generation and maintenances of turbulence, of supply, transfer, and dissipation of energy, and of local similarity are the foundation stones of turbulence theory.

PROBLEMS AND EXERCISES

1. Discuss the various types of instability mechanisms that might be operating in the atmosphere and their possible effects on atmospheric motions.
2. If you have available only the upper air velocity and temperature

soundings, how would you use this information to determine which layers of the atmosphere might be turbulent?

3. (a) What are the possible mechanisms of generation and maintenance of turbulence in the atmosphere?

 (b) What criteria may be used to determine if turbulence in a stably stratified layer could be maintained or if it would decay?

4. What are the characteristics of turbulence that distinguish it from a random wave motion?

5. Compare and contrast the time and ensemble averages. Under what conditions might the two types of averages be equivalent?

6. What is the possible range of turbulence intensities in the atmosphere and under what conditions might the extreme values occur?

7. Give all the components of the Reynolds stress and indicate which of these are really independent.

8. The following mean flow and turbulence measurements were made at a height of 22.6 m during the 1968 Kansas Field Program under different stability conditions:

Run no.	19	20	43	54	18	17	23
Ri	−1.89	−1.15	−0.54	−0.13	0.08	0.12	0.22
U (m sec^{-1})	4.89	6.20	8.06	9.66	7.49	5.50	5.02
σ_u (m sec^{-1})	0.83	1.04	1.27	1.02	0.73	0.45	0.20
σ_v (m sec^{-1})	1.02	1.07	1.26	0.91	0.56	0.36	0.17
σ_w (m sec^{-1})	0.65	0.73	0.73	0.64	0.48	0.28	0.10
σ_θ (K)	0.48	0.61	0.56	0.25	0.14	0.15	0.18
$\overline{uw}$ (m^2 sec^{-2})	−0.081	−0.111	−0.244	−0.214	−0.118	−0.043	−0.005
$\overline{\theta w}$ (m sec^{-1} K)	0.185	0.273	0.188	0.072	−0.029	−0.016	−0.005
$\overline{\theta u}$ (m sec^{-1} K)	−0.064	−0.081	−0.175	−0.129	0.068	0.039	0.023

 (a) Calculate and plot turbulence intensities as functions of Ri.

 (b) Calculate and plot correlation coefficients r_{uw}, $r_{w\theta}$, and $r_{u\theta}$ as functions of Ri.

 (c) Calculate the turbulent kinetic energy (TKE) for each run and plot the ratio $-\text{TKE}/\overline{uw}$ as a function of Ri.

 (d) What conclusion can you draw about the variation of the above turbulence quantities with stability?

Chapter 9 | Semiempirical Theories of Turbulence

9.1 MATHEMATICAL DESCRIPTION OF TURBULENT FLOWS

In Chapter 7 we introduced the equations of continuity, motion, and thermodynamic energy as mathematical expressions of the conservation of mass, momentum and heat in an elementary volume of fluid. These are applicable to laminar as well as turbulent flows. In the latter, all the variables and their temporal and spatial derivatives are highly irregular and rapidly varying functions of time and space. This essential property of turbulence makes all the terms in the equations significant, so that no further simplifications of them are feasible, aside from the Boussinesq approximations introduced in Chapter 7.

In particular, the instantaneous equations to the Boussinesq approximations for a thermally stratified turbulent flow, in a rotating frame of reference tied to the earth's surface, are

$$\frac{\partial \tilde{u}}{\partial x} + \frac{\partial \tilde{v}}{\partial y} + \frac{\partial \tilde{w}}{\partial z} = 0$$

$$\frac{\partial \tilde{u}}{\partial t} + \tilde{u}\frac{\partial \tilde{u}}{\partial x} + \tilde{v}\frac{\partial \tilde{u}}{\partial y} + \tilde{w}\frac{\partial \tilde{u}}{\partial z} = f\tilde{v} - \frac{1}{\rho_0}\frac{\partial \tilde{p}_1}{\partial x} + \nu\nabla^2\tilde{u}$$

$$\frac{\partial \tilde{v}}{\partial t} + \tilde{u}\frac{\partial \tilde{v}}{\partial x} + \tilde{v}\frac{\partial \tilde{v}}{\partial y} + \tilde{w}\frac{\partial \tilde{v}}{\partial z} = -f\tilde{u} - \frac{1}{\rho_0}\frac{\partial \tilde{p}_1}{\partial y} + \nu\nabla^2\tilde{v} \qquad (9.1)$$

$$\frac{\partial \tilde{w}}{\partial t} + \tilde{u}\frac{\partial \tilde{w}}{\partial x} + \tilde{v}\frac{\partial \tilde{w}}{\partial y} + \tilde{w}\frac{\partial \tilde{w}}{\partial z} = \frac{g}{T_0}\tilde{T}_1 - \frac{1}{\rho_0}\frac{\partial \tilde{p}_1}{\partial z} + \nu\nabla^2\tilde{w}$$

$$\frac{\partial \tilde{\theta}}{\partial t} + \tilde{u}\frac{\partial \tilde{\theta}}{\partial x} + \tilde{v}\frac{\partial \tilde{\theta}}{\partial y} + \tilde{w}\frac{\partial \tilde{\theta}}{\partial z} = \alpha_h\nabla^2\tilde{\theta}$$

Here, a tilde denotes an instantaneous variable which is the sum of its mean and fluctuative parts.

No general solution to this highly nonlinear system of equations is

123

known and there appears to be no hope of finding one through purely analytical methods.

9.1.1 FULL-TURBULENCE SIMULATION

In this age of computers, one may think of solving the Navier–Stokes equations using the brute force method of numerical integration on large computers. This has been called as full-turbulence simulation (FTS) and has been attempted for very low Reynolds number flows. The amount of computational effort to make a fundamental attack on most turbulent flows of practical interest is simply prohibitive and outside the range of even the most sophisticated supercomputers of today. The difficulty comes from the fact that the number of grid points required to resolve all the turbulent eddies in a flow is of the order of $(l/\eta)^3$, where l/η is the ratio of the largest to the smallest eddy scales. This number is simply mind boggling (10^{12}–10^{18}) for atmospheric turbulent flows and also too large to be handled for most other flows of practical interest. A rough estimate of the computer time required for a full-turbulence simulation of the daytime PBL using the latest Cray X-MP machine is millions of years! Still, certain low Reynolds number flows with $l/\eta \sim 10^2$ have been computed this way and, in the not too distant future, other turbulent flows of more practical interest might become amenable to the FTS approach, although still at a great cost.

9.1.2 LARGE-EDDY SIMULATION

A more feasible but less fundamental computational approach which has been used in meteorology and engineering fluid mechanics is large-eddy simulation (LES). This approach attempts to faithfully simulate only the scales of motion in a certain range (between the smallest grid size and the dimensions of simulated flow domain), while the unresolved subgrid scale motions are ignored altogether or are parameterized. The origins of the LES technique lie in the early global weather prediction and general circulation models in which the permissible grid spacing could hardly resolve large-scale atmospheric structures.

Large-eddy simulations of turbulent flows, including the PBL, started with the pioneering work of Deardorff (1970, 1973). The soundness of this modeling approach in simulating neutral and unstable PBLs, even in the presence of moist convection and clouds, has been amply demonstrated by subsequent studies of Deardorff and others. A pressing need is the extension of the LES to the nocturnal boundary layer, including the morning and evening transition periods, which are poorly understood. The main difficulty in simulating the stably stratified boundary layer is that the characteristic large-eddy scale (l) becomes too small and most of the energy transfer and other exchange processes are overly influenced or dominated by subgrid scale motions. A successful LES simulation of the NBL and transition processes should become possible in the near future, as more powerful next-generation computers become available.

The computational details of large-eddy simulation are outside the scope of this introductory text. It should suffice to say that LES is viewed as the most promising tool for future research in turbulence and is expected to provide a better understanding of the transport phenomena, well beyond the reach of semiempirical theories and most routine observations. Since it requires an investment of huge computer resources (there are also a few other limitations), LES would most likely remain a research tool. It would be very useful, of course, for generating numerical turbulence data for studying large-eddy structures, developing parameterizations for simpler models, and designing physical experiments (Wyngaard, 1984).

9.1.3 EQUATIONS OF MEAN MOTION

Full-turbulence simulation (FTS) is based on the "primitive" Navier–Stokes equations of motion. The large-eddy simulation utilizes the same equations, but averaged over the finite grid volume. Numerical integrations of both systems yield highly irregular (turbulent) variables as functions of time and space. Average statistics, such as means, variances, and higher moments, are then calculated from this "simulated turbulence" data. In most applications, however, only the average statistics are needed and how they come about from turbulent fluctuations is usually of little or no concern. One may ask "Why not use the equations of mean motion to compute the desired average statistics?" This question was first addressed by Osborne Reynolds toward the end of the nineteenth century. He also suggested some simple averaging rules or conditions and derived what came to be known as the Reynolds-averaged equations.

If f and g are two dependent variables or functions of variables with mean values $\bar{f}$ and $\bar{g}$, and c is a constant, the Reynolds averaging conditions are

$$\overline{f + g} = \bar{f} + \bar{g}$$
$$\overline{cf} = c\bar{f}; \qquad \overline{\bar{f}g} = \bar{f}\bar{g} \qquad (9.2)$$
$$\overline{\partial f/\partial s} = \partial \bar{f}/\partial s; \qquad \int \overline{f\,ds} = \int \bar{f}\,ds$$

where $s = x, y, z$, or t. These conditions are used in deriving the equations for mean variables from those of the instantaneous variables. Since, these are strictly satisfied only by ensemble averaging, this type of averaging is generally implied in theory. The time and space averages often used in practice can satisfy the Reynolds averaging rules only under certain idealized conditions (e.g., stationarity and homogeneity of flow), which were discussed in Chapter 8.

The usual procedure for deriving Reynolds-averaged equations is to substitute in Eq. (9.1) $\tilde{u} = U + u$, $\tilde{v} = V + v$, etc., and take their averages using the Reynolds averaging conditions [Eq. (9.2)]. Consider first the continuity equation

$$\partial(U + u)/\partial x + \partial(V + v)/\partial y + \partial(W + w)/\partial z = 0 \qquad (9.3)$$

which, after averaging, gives

$$\partial U/\partial x + \partial V/\partial y + \partial W/\partial z = 0 \qquad (9.4)$$

Subtracting Eq. (9.4) from (9.3), one obtains the continuity equation for the fluctuating motion

$$\partial u/\partial x + \partial v/\partial y + \partial w/\partial z = 0 \qquad (9.5)$$

Thus, the form of continuity equation remains the same for instantaneous, mean, and fluctuating motions. This is not the case, however, for the equations of conservation of momentum and heat, because of the presence of nonlinear advection terms in those equations.

Let us consider, for example, the advection terms in the instantaneous equation of the conservation of heat

$$\tilde{a}_\theta = \tilde{u}(\partial\tilde{\theta}/\partial x) + \tilde{v}(\partial\tilde{\theta}/\partial y) + \tilde{w}(\partial\tilde{\theta}/\partial z) \qquad (9.6a)$$

which, after utilizing the continuity equation, can also be written in the form

$$\tilde{a}_\theta = (\partial/\partial x)(\tilde{u}\tilde{\theta}) + (\partial/\partial y)(\tilde{v}\tilde{\theta}) + (\partial/\partial z)(\tilde{w}\tilde{\theta}) \qquad (9.6b)$$

Expressing variables as sums of their mean and fluctuating parts in Eq. (9.6b) and averaging, one obtains the Reynolds-averaged advection terms

$$A_\theta = \frac{\partial}{\partial x}(U\Theta) + \frac{\partial}{\partial y}(V\Theta) + \frac{\partial}{\partial z}(W\Theta) \qquad (9.7a)$$
$$+ \frac{\partial}{\partial x}(\overline{u\theta}) + \frac{\partial}{\partial y}(\overline{v\theta}) + \frac{\partial}{\partial z}(\overline{w\theta})$$

or, after using the mean continuity equation,

$$A_\theta = U\frac{\partial\Theta}{\partial x} + V\frac{\partial\Theta}{\partial y} + W\frac{\partial\Theta}{\partial z} + \frac{\partial}{\partial x}(\overline{u\theta}) + \frac{\partial}{\partial y}(\overline{v\theta}) + \frac{\partial}{\partial z}(\overline{w\theta}) \qquad (9.7b)$$

Thus, upon averaging, the nonlinear advection terms yield not only the terms which may be interpreted as advection or transport by mean flow, but also several additional terms involving covariances or turbulent fluxes. These latter terms are the spatial gradients (divergence) of turbulent transports.

Following the above procedure with each of the equalities given in Eq. (9.1), one obtains the Reynolds-averaged equations for the conservation of mass, momentum, and heat:

$$\frac{\partial U}{\partial x} + \frac{\partial V}{\partial y} + \frac{\partial W}{\partial z} = 0$$

$$\frac{\partial U}{\partial t} + U\frac{\partial U}{\partial x} + V\frac{\partial U}{\partial y} + W\frac{\partial U}{\partial z} = fV - \frac{1}{\rho_0}\frac{\partial P_1}{\partial x} + \nu\nabla^2 U$$
$$- \left(\frac{\partial \overline{u^2}}{\partial x} + \frac{\partial \overline{uv}}{\partial y} + \frac{\partial \overline{uw}}{\partial z}\right)$$

$$\frac{\partial V}{\partial t} + U\frac{\partial V}{\partial x} + V\frac{\partial V}{\partial y} + W\frac{\partial V}{\partial z} = -fU - \frac{1}{\rho_0}\frac{\partial P_1}{\partial y} + \nu\nabla^2 V \qquad (9.8)$$
$$- \left(\frac{\partial \overline{uv}}{\partial x} + \frac{\partial \overline{v^2}}{\partial y} + \frac{\partial \overline{vw}}{\partial z}\right)$$

$$\frac{\partial W}{\partial t} + U\frac{\partial W}{\partial x} + V\frac{\partial W}{\partial y} + W\frac{\partial W}{\partial z} = \frac{g}{T_0}T_1 - \frac{1}{\rho_0}\frac{\partial P_1}{\partial z} + \nu\nabla^2 W$$
$$- \left(\frac{\partial \overline{wu}}{\partial x} + \frac{\partial \overline{wv}}{\partial y} + \frac{\partial \overline{w^2}}{\partial z}\right)$$

$$\frac{\partial \Theta}{\partial t} + U\frac{\partial \Theta}{\partial x} + V\frac{\partial \Theta}{\partial y} + W\frac{\partial \Theta}{\partial z} = \alpha_h\nabla^2\Theta - \left(\frac{\partial \overline{u\theta}}{\partial x} + \frac{\partial \overline{v\theta}}{\partial y} + \frac{\partial \overline{w\theta}}{\partial z}\right)$$

When these equations are compared with the corresponding instantaneous equations [Eq. (9.1)], one finds that most of the terms (except for the turbulent transport terms) are similar and can be interpreted in the same way. However, there are several fundamental differences between these two sets of equations, which forces us to treat them in entirely different ways. While Eq. (9.1) deals with instantaneous variables varying rapidly and irregularly in time and space, Eq. (9.8) deals with mean variables which are comparatively well behaved and vary only slowly and smoothly. While all the terms in the former equation set may be significant and cannot be ignored *a priori*, the mean flow equations can be greatly simplified by neglecting the molecular diffusion terms outside of possible viscous sublayers and also other terms on the basis of certain boundary layer approximations and considerations of stationarity and horizontal homogeneity, whenever applicable. For example, it is easy to show that for a horizontally homogeneous and stationary PBL the equations of mean flow reduce to

$$-f(V - V_g) = -\partial \overline{uw}/\partial z$$
$$f(U - U_g) = -\partial \overline{vw}/\partial z$$

$$(9.9)$$

An important consequence of averaging the equations of motion is the appearance of turbulent flux-divergence terms which contain yet unknown variances and covariances. Note that in Eq. (9.8) there are many more unknowns than the number of equations. While the number of extra unknowns may be reduced to one or two in certain simple flow situations, the Reynolds-averaged equations remain essentially unclosed and, hence, unsolvable. This so-called closure problem of turbulence has been a major stumbling block in developing a rigorous and general theory of turbulence. It is a consequence of the nonlinearity of the original (instantaneous) equations of motion. Many semiempirical theories and models have been proposed to get around the closure problem, but none of them has proved to be entirely satisfactory. Here, we will briefly discuss only the simplest gradient-transport hypotheses or relations, which are still widely used in micrometeorology.

9.2   GRADIENT-TRANSPORT THEORIES

In order to close the set of equations given by Eq. (9.8) or its simplified version for a given flow situation, the variances and covariances must either be specified in terms of other variables or additional equations must be developed for the same. In the latter approach, the closure problem is only shifted to a higher level in the hierarchy of equations that can be developed. We will not discuss here the so-called higher order closure schemes or models that have been developed in recent years and that have been found to have their own limitations and problems. The older and more widely used approach has been based on the assumed (hypothetical) analogy between molecular and turbulent transfers. It is called the gradient-transport approach, because turbulent transports or fluxes are sought to be related to the appropriate gradients of mean variables (velocity, temperature, etc.). Several different hypotheses have been used in developing such relationships.

9.2.1 EDDY VISCOSITY (DIFFUSIVITY) HYPOTHESIS

In analogy with Newton's law of molecular viscosity [see Eq. (7.1), Chapter 7], J. Boussinesq in 1877 proposed that the turbulent shear stress in the direction of flow may be expressed as

$$\tau = \rho K_m (\partial U / \partial z) \tag{9.10}$$

in which K_m is called the eddy exchange coefficient of momentum or simply the eddy viscosity, which is analogous to the molecular kinematic viscosity ν. In analogy with the more general constitutive relations [Eq. (7.2)], one can also generalize Eq. (9.10) to express the various Reynolds stress components in terms of mean gradients. In particular, when the mean gradients in the x and y directions can be neglected in comparison to those in the z direction (the usual boundary layer approximation), we have the eddy viscosity relations for the vertical fluxes of momentum

$$\overline{uw} = -K_m (\partial U / \partial z)$$
$$\overline{vw} = -K_m (\partial V / \partial z) \tag{9.11}$$

Similar relations have been proposed for the turbulent fluxes of heat, water vapor, and other transferrable constituents (e.g., pollutants), which are analogous to Fourier's and Fick's laws of molecular diffusion of heat and mass. Those frequently used in micrometerology are the relations for the vertical fluxes of heat $(\overline{\theta w})$ and water vapor $(\overline{qw})$

$$\overline{\theta w} = -K_h (\partial \Theta / \partial z)$$
$$\overline{qw} = -K_w (\partial Q / \partial z) \tag{9.12}$$

in which K_h and K_w are called the exchange coefficients or eddy diffusivities of heat and water vapor, respectively, and Q and q denote the mean and fluctuating parts of specific humidity.

It should be recognized that the above gradient-transport relations are not the expressions of any sound physical laws in the same sense that their molecular counterparts are. These are not based on any rigorous theory, but only on an intuitive assumption of similarity or analogy between molecular and turbulent transfers. Under ordinary circumstances, one would expect heat to flow from warmer to colder regions, roughly in proportion to the temperature gradient. Similarly, momentum and mass transfers may be expected to be proportional to and down the mean gradients. However, these expectations are not always borne out by experimental data in turbulent flows, including the atmospheric boundary layer.

The analogy between molecular and turbulent transfers has subsequently been found to be very weak and qualitative only. Eddy diffusivities, as determined from their defining relations Eqs. (9.11) and (1.12), are usually several orders of magnitude larger than their molecular counterparts, indicating the dominance of turbulent mixing over molecular exchanges. More importantly, eddy diffusivities cannot simply be regarded as fluid properties; these are actually turbulence or flow properties, which can vary widely from one flow to another and from one region to another in the same flow. Eddy diffusivities show no apparent dependence on molecular properties, such as mass density, temperature, etc., and have nothing in common with molecular diffusivities, except for the same dimensions.

In spite of the above limitations of the implied analogy between molecular and turbulent diffusion, Eqs. (9.10) and (9.12) need not be restrictive, since they replace only one set of unknowns (fluxes) for another (eddy diffusivities). Some restrictions are imposed, however, when one assumes that eddy diffusivities depend on the coordinates and flow parameters in some definite manner. This then constitutes a semiempirical theory that is based on a hypothesis and is subject to experimental verification. The simplest assumption, which Boussinesq proposed originally, is that eddy diffusivities are constants for the whole flow. It turns out that this assumption works well in free turbulent flows such as jets, wakes, and mixing layers, away from any boundaries, and is often used in the free atmosphere. But when applied to boundary layers and channel flows, it leads to incorrect results. In general, the assumption of constant eddy diffusivity is quite inapplicable near a rigid surface. But here other reasonable hypotheses can be made regarding the variation of eddy diffusivity with distance from the surface. For example, a linear distribution of K_m in the neutral surface layer works quite well. Suggested modifications of the K distribution in thermally stratified conditions are usually based on other theoretical considerations and empirical data, which will be discussed later.

There are other limitations of the K theory which we have not mentioned so far. The basic notion of down-gradient transport implied in the theory may be questioned. There are practical situations when turbulent fluxes are in no way related to the local gradients. For example, in a convective mixed layer the potential temperature gradient becomes near zero or slightly positive, while the heat is transported upward in significant amounts. This would imply infinite or even negative values of K_h, indicating that K theory becomes invalid in this case. Even in other situations, the specification of eddy diffusivities in a rational manner is quite difficult, if not impossible. Still, the theory is quite useful and is widely used in practice.

9.2.2 MIXING-LENGTH HYPOTHESIS

In an attempt to specify eddy viscosity as a function of geometry and flow parameters, L. Prandtl in 1925 further extended the molecular analogy by ascribing a hypothetical mechanism for turbulent mixing. According to the kinetic theory of gases, momentum and other properties are transferred when molecules collide with each other. The theory leads to an expression of molecular viscosity as a product of mean molecular velocity and the mean free path length (the average distance traveled by molecules before collision). Prandtl hypothesized a similar mechanism of transfer in turbulent flows by assuming that eddies or "blobs" of fluid (analogous to molecules) break away from the main body of the fluid and travel a certain distance, called the mixing length (analogous to free path length), before they mix suddenly with the new environment. If the velocity, temperature, and other properties of a blob or parcel are different from those of the environment with which the parcel mixes, fluctuations in these properties would be expected to occur as a result of the exchanges of momentum, heat, etc. If such eddy motions occur more or less randomly in all directions, it can be easily shown that net (average) exchanges of momentum, heat, etc., will occur only in the direction of decreasing velocity, temperature, etc.

In order to illustrate the above mechanism for the generation of turbulent fluctuations and their covariances (fluxes), we consider the usual case of increasing mean velocity with height in the lower atmosphere (see Fig. 9.1).

The longitudinal velocity fluctuations at a level z may be assumed to occur as a result of mixing with the environment, at this level, of fluid parcels arriving from different levels above and below. For example, a parcel arriving from below (level $z - l$) will give rise to a negative fluctuation at level z of magnitude

$$u = U(z - l) - U(z) \simeq -l(\partial U/\partial z) \tag{9.13}$$

associated with its positive vertical velocity (fluctuation) w. The last approximation is based on the assumption of a linear velocity profile over the mixing-length l, which is considered here a fluctuating quantity with positive values for the upward motions and negative for the downward motions of the parcel. Considering the action of many parcels arriving at z and taking the average, we obtain an expression of the momentum flux

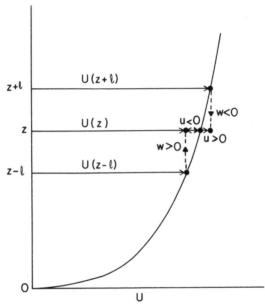

Fig. 9.1 Schematic of mean velocity profile in the lower atmosphere and expected correlations between the longitudinal and vertical velocity fluctuations due to fluid parcels coming from above or below before mixing with the surrounding fluid.

$$\overline{uw} = -\overline{lw}(\partial U/\partial z) \tag{9.14}$$

which is not very helpful, because there is no direct way of measuring l.

Another mixing-length expression for $\overline{uw}$ was obtained by Prandtl, after assuming that in a turbulent flow the velocity fluctuations in all directions are of the same order of magnitude and are related to each other, so that

$$w \sim -u \simeq l(\partial U/\partial z) \tag{9.15}$$

From Eqs. (9.13) and (9.15) one obtains

$$\overline{uw} \sim -\overline{l^2}(\partial U/\partial z)^2$$

or Prandtl's original mixing-length relation

$$\overline{uw} = -l_m^2(|\partial U/\partial z|)(\partial U/\partial z) \tag{9.16}$$

in which l_m is a mean mixing length, which is proportional to the root-mean-square value of the fluctuating length l, and the absolute value of the velocity gradient is introduced to ensure that the momentum flux is down the gradient.

Equation (9.16) constitutes a closure hypothesis, if the mixing length l_m can be prescribed as a function of the flow geometry and, possibly, other flow properties. If large eddies are mainly responsible for momentum and other exchanges in a turbulent flow it can be argued that l_m should be directly related to the characteristic large-eddy length scale. Some investigators make no distinction between the two, although, strictly speaking, a proportionality coefficient of the order of one is usually involved. In free turbulent flows and in the outer parts of boundary layer and channel flows, where large-eddy size may not have significant spatial variations, the assumption of constant mixing length ($l_m = l_0$) is found to be reasonable. In surface or wall layers, the large-eddy size, at least normal to the surface, varies roughly in proportion to the distance from the surface, so that mixing length is also expected to be proportional to the distance ($l_m \sim z$). In the atmospheric boundary layer, the mixing-length distribution is further expected to depend on thermal stratification and boundary layer thickness, and will be considered later.

The mixing-length hypothesis may be used to specify eddy viscosity. From a comparison of Eqs. (9.11) and (9.16), it is obvious that

$$K_m = l_m^2(|\partial U/\partial z|) \tag{9.17}$$

An alternative specification is obtained from a comparison of Eqs. (9.11) and (9.14)

$$K_m = c_m l_m \sigma_w \tag{9.18}$$

where c_m is a constant. Other parametric relations of eddy viscosity involving the product of a characteristic length scale and a turbulence velocity scale have been proposed in the literature. Similar relations are given for K_h and K_w.

9.3 DIMENSIONAL ANALYSIS AND SIMILARITY THEORIES

Dimensional analysis and similarity considerations are extensively used in micrometeorology, as well as in other areas of science and engineering. Therefore, these analytical methods are briefly discussed in this section, while their applications in micrometeorology will be given in later chapters.

9.3.1 DIMENSIONAL ANALYSIS

Dimensional analysis is a simple but powerful method of investigating a variety of scientific phenomena and establishing useful relationships between the various quantities or parameters, based on their dimensions.

One can define a set of fundamental dimensions, such as length $[L]$, time $[T]$, mass $[M]$, etc., and express the dimensions of all the quantities involved in terms of these fundamental dimensions. A representation of the dimensions of a quantity or a parameter in terms of fundamental dimensions constitutes a dimensional formula, e.g., the dimensional formula for fluid viscosity is $[\mu] = [ML^{-1}T^{-1}]$. If the exponents in the dimensional formula are all zero, the parameter under consideration is dimensionless. One can form dimensionless parameters from appropriate combinations of dimensional quantities; e.g., the Reynolds number $Re = VL\rho/\mu$ is a dimensionless combination of fluid velocity V, the characteristic length scale L, density ρ, and viscosity μ.

Dimensionless groups or parameters are of special significance in any dimensional analysis in which the main objective is to seek certain functional relationships between the various dimensionless parameters. There are several reasons for considering dimensionless groups instead of dimensional quantities or variables. First, mathematical expressions of fundamental physical laws are dimensionally homogeneous (i.e., all the terms in an expression or equation have the same dimensions) and can be written in dimensionless forms simply by an appropriate choice of scales for normalizing the various quantities. Second, dimensionless relations represented in mathematical or graphical form are independent of the system of units used and they facilitate comparisons between data obtained by different investigators at different locations and times. Third, and, perhaps, the most important reason for working with dimensionless parameters is that nondimensionalization always reduces the number of parameters that are involved in a functional relationship. This follows from the well-known Buckingham pi theorem, which states that if m quantities $(Q_1, Q_2, ..., Q_m)$, involving n fundamental dimensions, form a dimensionally homogeneous equation, the relationship can always be expressed in terms of $m - n$ independent dimensionless groups $(\Pi_1, \Pi_2, ..., \Pi_{m-n})$ made of the original m quantities. Thus, the dimensional functional relationship

$$f(Q_1, Q_2, ..., Q_m) = 0 \qquad (9.19)$$

is equivalent to the dimensionless relation

$$F(\Pi_1, \Pi_2, ..., \Pi_{m-n}) = 0 \qquad (9.20)$$

or, alternatively,

$$\Pi_1 = F_1(\Pi_2, \Pi_3, ..., \Pi_{m-n}) \qquad (9.21)$$

In particular, when only one dimensionless group can be formed out of all the quantities, i.e., when $m - n = 1$, that group must be a constant, since it cannot be a function of any other parameter. Dimensional analysis does not give actual forms of the functions F, F_1, etc., or values of any dimensionless constants that might result from the analysis. This must be done by other means, such as further theoretical considerations and experimental observations. It is common practice to follow dimensional analysis by a systematic experimental study of the phenomenon to be investigated.

9.3.2 AN ILLUSTRATIVE EXAMPLE

In order to illustrate the method and usefulness of dimensional analysis, let us consider the possible relationship between the mean potential temperature gradient $(\partial\Theta/\partial z)$, the height (z) above a uniform heated surface, the surface heat flux (H_0), the buoyancy parameter (g/T_0) which appears in the expressions for static stability and buoyant acceleration, and the relevant fluid properties $(\rho$ and $c_p)$ in the near-surface layer when free convection dominates any mechanical mixing (this latter condition permits dropping of all shear-related parameters from consideration). To establish a functional relationship in the dimensional form

$$f(\partial\Theta/\partial z, H_0, g/T_0, z, \rho, c_p) = 0 \tag{9.22}$$

would require extensive observations of temperature as a function of height and surface heat flux at different times and locations (to represent different types of surfaces and radiative regimes). If the relationship is to be further generalized to other fluids, laboratory experiments using these different fluids will also be necessary. Considerable simplification can be achieved, however, if ρ and c_p are combined with H_0 to form a kinematic heat flux parameter $H_0/\rho c_p$, so that Eq. (9.22) can be written as

$$F(\partial\Theta/\partial z, H_0/\rho c_p, g/T_0, z) = 0 \tag{9.23}$$

If we now use the method of dimensional analysis, realizing that only one dimensionless group can be formed from the above quantities, we obtain

$$(\partial\Theta/\partial z)(H_0/\rho c_p)^{-2/3}(g/T_0)^{1/3}z^{4/3} = C \tag{9.24}$$

in which the left-hand side is the dimensionless group that is predicted to be a constant. The value of the constant C can be determined from only one carefully conducted experiment, although a thorough experimental verification of the above relationship might require more extensive observations.

The desired dimensionless group from a given number of quantities can often be formed merely by inspection. A more formal and general approach would be to write and solve a system of algebraic equations for the exponents of various quantities involved in the dimensionless group. For example, the dimensionless group formed out of all the parameters in Eq. (9.23) may be assumed as

$$\Pi_1 = (\partial\Theta/\partial z)(H_0/\rho c_p)^a(g/T_0)^b z^c \tag{9.25}$$

in which we have arbitrarily assigned a value of unity to one of the indices (here, the exponent of $\partial\Theta/\partial z$), because any arbitrary power of a dimensionless quantity is also dimensionless. Writing Eq. (9.25) in terms of our chosen fundamental dimensions (length, time, and temperature), we have

$$[L^0 T^0 K^0] = [KL^{-1}][KLT^{-1}]^a[LT^{-2} K^{-1}]^b[L]^c$$

from which we obtain the equations

$$0 = -1 + a + b + c$$
$$0 = -a - 2b \tag{9.26}$$
$$0 = 1 + a - b$$

whose solution gives $a = -2/3$, $b = 1/3$, and $c = 4/3$. Substituting these values in Eq. (9.25) and equating the only dimensionless group to a constant, then, yields Eq. (9.24).

Another approach is to first formulate the characteristic scales of length, velocity, etc., from combinations of independent variables and then use these scales to normalize the dependent variables. In the case of multiples scales, their ratios form the independent dimensionless groups. In the above example of temperature distribution over a heated surface, if we choose $\partial\Theta/\partial z$ as the dependent variable and the remaining quantities as independent variables, the following scales can be formulated out of the latter:

$$z \qquad\qquad\qquad \text{length scale}$$

$$\theta_f \equiv \left(\frac{H_0}{\rho c_p}\right)^{2/3} \left(\frac{g}{T_0}\right)^{-1/3} z^{-1/3} \qquad \text{temperature scale} \tag{9.27}$$

$$u_f \equiv \left(\frac{H_0}{\rho c_p}\frac{g}{T_0} z\right)^{1/3} \qquad \text{velocity scale}$$

Then, the appropriate dimensionless group involving the dependent variable is $(z/\theta_f)(\partial\Theta/\partial z)$, which must be a constant, since no other independent dimensionless groups can be formed from independent variables. This procedure also leads to Eq. (9.24); it is found to be more convenient to use when a host of dependent variables are functions of the same set of independent variables. For example, standard deviations of temperature and vertical velocity fluctuations in the free convective surface layer are given by

$$\sigma_\theta/\theta_f = C_\theta$$
$$\sigma_w/u_f = C_w$$

(9.28)

or, after substituting from Eq. (9.27),

$$\sigma_\theta = C_\theta(H_0/\rho c_p)^{2/3}(gz/T_0)^{-1/3}$$
$$\sigma_w = C_w[(H_0/\rho c_p)(g/T_0)z]^{1/3}$$

(9.29)

Equations (9.29) have proved to be very useful relations for the atmospheric surface layer under daytime unstable conditions and are supported by many observations (see, e.g., Monin and Yaglom, 1971, Chap. 5; Wyngaard, 1973) from which $C_\theta \simeq 1.3$ and $C_w \simeq 1.4$.

9.3.3 SIMILARITY THEORY

The pi theorem and dimensional analysis discussed above are merely mathematical formalisms and do not deal with the physics of the problem. In order to use them, one has to know or correctly guess, using physical intuition and available experimental information, the relevant quantities involved in any desired mathematical or empirical relationship. This constitutes a plausible hypothesis on the dependence of a desired variable on a number of independent variables or physical parameters. This first step is the most crucial one in the development of a plausible similarity theory based on dimensional analysis. On the one hand, one cannot ignore any of the important quantities on which the phenomenon or variable under investigation really depends, because this might lead to completely wrong and unphysical relationships. On the other hand, if unnecessary and irrelevant quantities are included in the original similarity hypothesis, they will complicate the analysis and make empirical determination of the various functional relationships extremely difficult, if not impossible.

Going back to our previous example, if we had ignored any one of the quantities in Eq. (9.23), we could not have formed even one dimensionless group, which would be indicative of an important omission. On the other hand, if we had added another variable, e.g., the boundary layer height h, to the list in Eq. (9.23), we would have, according to the pi theorem, two independent dimensionless groups and the predicted functional relationship would be

$$\frac{\partial\Theta}{\partial z}\left(\frac{H_0}{\rho c_p}\right)^{-2/3}\left(\frac{g}{T_0}\right)^{1/3}z^{4/3} = F(z/h)$$

(9.30)

Subsequent analysis of experimental data might indicate, however, that the functional dependence of the left-hand side on z/h is very weak or nonexistent and, hence, the irrelevance of h in the original hypothesis. The inclusion of h would be quite justified and even necessary, however, if we were to investigate the turbulence structure of the mixed layer. With this, the relevant mixed-layer similarity scales are

$$h \qquad\qquad \text{length scale}$$

$$T_* \equiv \left(\frac{H_0}{\rho c_p}\right)^{2/3} \left(\frac{gh}{T_0}\right)^{-1/3} \qquad \text{temperature scale} \qquad (9.31)$$

$$W_* = \left(\frac{H_0}{\rho c_p} \frac{g}{T_0} h\right)^{1/3} \qquad \text{velocity scale}$$

The corresponding similarity predictions are that the dimensionless structure parameters, σ_θ/T_*, σ_u/W_*, σ_w/W_*, etc., must be some unique function of z/h. This mixed-layer similarity theory has proved to be very useful in describing turbulence and diffusion in the convective boundary layer.

As the number of independent variables and parameters is increased in the original hypothesis, not only the number of independent dimensionless groups increases, but also the possible combinations of variables in forming such groups become large. The possibility of experimentally determining their functional relationships becomes increasingly remote as the number of dimensionless groups increases beyond two or three. Therefore, it is always desirable to keep the number of independent variables to a minimum, consistent with physics. Sometimes, it may require breaking the domain of the problem or phenomenon under investigation into several smaller subdomains, so that simpler similarity hypotheses can be formulated for each of them separately. For example, the atmospheric PBL is usually divided into a surface layer and an outer layer for dimensional analysis and similarity considerations. One can further divide the flow regime according to stability (e.g., stable, neutral, and convective PBLs) and baroclinity (e.g., barotropic and baroclinic PBLs) and use different similarity hypotheses for different conditions. A generalized PBL similarity theory will have to include all the possible factors influencing the PBL under the whole range of conditions encountered and, hence, will be too unwieldy for practical use.

The functional relationships between dimensionless groups that come out of dimensional analysis are sometimes referred to as similarity relations, because they express the conditions under which two or more flow regimes would be similar. Equality of certain dimensionless parameters is required for similarity. For the same reason, the original hypothesis preceding dimensional analysis is called a similarity hypothesis and the analysis based on the same is referred to as a similarity analysis or theory. A variety of similarity theories have been used in micrometeorology; a few of these will be discussed in the following chapters.

9.4 APPLICATIONS

The theories and models of turbulence discussed in this chapter are widely used in micrometeorology. Some of the specific applications are as follows:

- Calculating the mean structure (e.g., vertical profiles of mean velocity and temperature) of the PBL
- Calculating the turbulence structure (e.g., profiles of fluxes, variances of fluctuations, and scales of turbulence) of the PBL
- Providing plausible theoretical explanations for turbulent exchange and mixing processes in the PBL
- Providing suitable frameworks for analyzing and comparing micrometeorological data from different sites
- Suggesting simple methods of estimating turbulent fluxes from mean profile observations

PROBLEMS AND EXERCISES

1. Compare and contrast the instantaneous and the Reynolds-averaged equations of motion for the PBL.
2. (a) What boundary layer approximations are often used for simplifying the Reynolds-averaged equations of motion for the PBL?
   (b) What are the other simplifying assumptions commonly used in micrometeorology and the conditions in which they may not be valid?
3. Show, step by step, how Eq. (9.8) is reduced to Eq. (9.9) for a horizontally homogeneous and stationary PBL.
4. (a) Write down the mean thermodynamic energy equation for a horizontally homogeneous PBL and discuss the rationale for retaining the time-tendency term in the same.
   (b) Describe a method for estimating the diurnal variation of surface heat flux from hourly soundings of temperature in the PBL.
5. Show that with the assumption of a constant eddy viscosity the equations of mean motion in the PBL become similar to those for the

laminar Ekman layer and, hence, discuss some of the limitations of the above assumption.

6. If the surface layer turbulence under neutral stability conditions is characterized by a large-eddy length scale l, which is proportional to height, and a velocity scale u_*, which is independent of height, suggest the plausible expressions of mixing length and eddy viscosity in this layer.

7. A similarity hypothesis proposed by von Karman states that mixing length in a turbulent shear flow depends only on the first and second spatial derivatives of mean velocity $(\partial U/\partial z, \partial^2 U/\partial z^2)$ in the direction of shear. Suggest an expression for mixing length which is consistent with the above similarity hypothesis.

8. (a) In a neutral, barotropic PBL, ageostrophic velocity components $(U - U_g$ and $V - V_g)$ may be assumed to depend only on the height z above the surface, the friction velocity scale u_*, and the Coriolis parameter f. On the basis of dimensional analysis suggest the appropriate expressions for velocity distribution and the PBL height h.

   (b) If the PBL height in Problem 8(a) above was determined by a low-level inversion, with its base at z_i, what should be the corresponding form of dimensionless velocity distribution?

9. (a) In a convective boundary layer during the midday period when $T_0 = 300$ K, $H_0 = 500$ W m^{-2}, and $h = 1500$ m, calculate and compare the standard deviations of velocity and temperature fluctuations at the heights of 10 and 100 m.

   (b) Show that, in the free convective surface layer,
   $$\sigma_\theta/T_* = C_\theta(z/h)^{-1/3}; \qquad \sigma_w/W_* = C_w(z/h)^{1/3}$$

Chapter 10 | Neutral Boundary Layers

10.1 VELOCITY-PROFILE LAWS

Strictly neutral stability conditions are rarely encountered in the atmosphere. However, during overcast skies and strong surface geostrophic winds, the atmospheric boundary layer may be considered near-neutral, and simpler theoretical and semiempirical approaches developed for neutral boundary layers by fluid dynamists and engineers can be used in micrometeorology. One must recognize, however, that, unlike in unidirectional flat-plate boundary layer and channel flows, the wind direction in the PBL changes with height in response to Coriolis or rotational effects. Therefore, wind distribution in the PBL is expressed in terms of either wind speed and direction, or the two horizontal components of velocity (see, e.g., Figs. 6.5–6.8).

10.1.1 THE POWER-LAW PROFILE

Measured velocity distributions in flat-plate boundary layer and channel flows can be approximately represented by a power law of the form

$$U/U_h = (z/h)^m \tag{10.1}$$

which was originally suggested by L. Prandtl with an exponent $m = 1/7$ for smooth surfaces. Here, h is the boundary layer thickness or half-channel depth. Since, wind speed does not increase monotonically with height up to the top of the PBL, a slightly modified version of Eq. (10.1) is used in micrometeorology:

$$U/U_r = (z/z_r)^m \tag{10.2}$$

where U_r is the wind speed at a reference height z_r, which is smaller than or equal to the height of wind speed maximum; a standard reference height of 10 m is commonly used.

The power-law profile does not have a sound theoretical basis, but

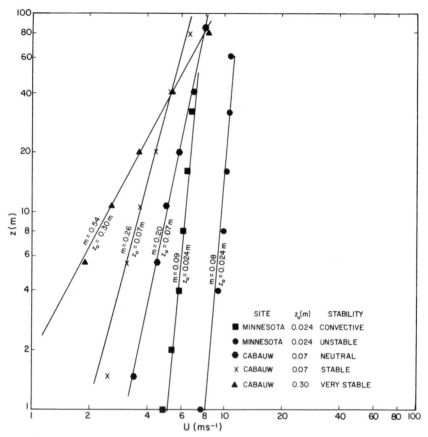

Fig. 10.1 Comparison of observed wind speed profiles at different sites (z_0 is a measure of the surface roughness) under different stability conditions with the power-law profile. [Data from Izumi and Caughey (1976).]

frequently it provides a reasonable fit to the observed velocity profiles in the lower part of the PBL, as shown in Fig. 10.1. The exponent m is found to depend on both the surface roughness and stability. Under near-neutral conditions, the values of m range from 0.10 for smooth water, snow, and ice surfaces to about 0.40 for well-developed urban areas. Figure 10.2 shows the dependence of m on the surface roughness parameter (length) z_0, which will be defined later. The exponent m also increases with increasing stability and approaches one (corresponding to a linear profile) under very stable conditions. The value of the exponent may also depend, to some extent, on the height range over which the power law is fitted to the observed profile.

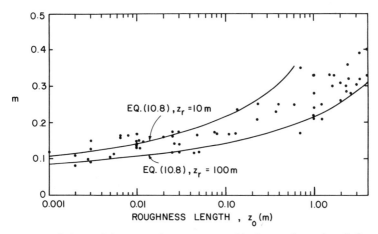

Fig. 10.2 Variations of the power-law exponent with the roughness length for near-neutral conditions. [Data reprinted with permission from *Atmospheric Environment,* Vol. 9, J. C. Counihan, Adiabatic atmospheric boundary layers: A review and analysis of data from the period 1880–1972, Copyright (1975), Pergamon Journals Ltd.]

The power-law velocity profile implies a power-law eddy viscosity (K_m) distribution in the lower part of the boundary layer, in which the momentum flux may be assumed to remain nearly constant with height, i.e., in the constant stress layer. It is easy to show that

$$K_m/K_{mr} = (z/z_r)^n \qquad (10.3)$$

with the exponent $n = 1 - m$, is consistent with Eq. (10.2) in the surface layer. Equations (10.2) and (10.3) are called conjugate power laws and have been extensively used in theoretical formulations of atmospheric diffusion, including transfers of heat and water vapor from extensive uniform surfaces (see Sutton, 1953). In such formulations eddy diffusivities of heat and mass are assumed to be equal or proportional to eddy viscosity, and thermodynamic energy and diffusion equations are solved for prescribed velocity and eddy diffusivity profiles in the above manner. When Eqs. (10.2) and (10.3) are used above the constant stress layer, the conjugate relationship $n = 1 - m$ need not be satisfied; this relationship may be too restrictive even in the surface layer.

10.1.2 THE LOGARITHMIC PROFILE LAW

Let us focus our attention now to the netural surface layer over a flat and uniform surface in which Coriolis effects can be ignored and the momentum flux may be considered constant, independent of height. Furthermore, we also exclude from our consideration any viscous sublayer that may exist over a smooth surface, or the canopy or roughness sublayer in which the flow is very likely to be disturbed by individual roughness elements. For the remaining fully turbulent, horizontally homogeneous surface layer, a simple similarity hypothesis can be used to obtain the velocity distribution, namely, the wind shear $\partial U/\partial z$ is only dependent on the height z above the surface (more appropriately, above a suitable reference plane near the surface), the surface drag, and the fluid density, i.e.,

$$\partial U/\partial z = f(z, \tau_0, \rho) \tag{10.4}$$

An implied assumption in this similarity hypothesis is that the influence of other possible parameters, such as the surface roughness, horizontal pressure gradients (geostrophic winds), and the PBL height, is fully accounted for in τ_0, which then determines the velocity gradients in the surface layer.

The only characteristic velocity scale given by the above surface layer similarity hypothesis is the so-called friction velocity $u_* \equiv (\tau_0/\rho)^{1/2}$, and the only characteristic length scale is z. Then, from dimensional analysis it follows that the dimensionless wind shear

$$(z/u_*)(\partial U/\partial z) = \text{const.} = 1/k \tag{10.5}$$

where k is called von Karman's constant.

The above similarity relation has been verified by many observed velocity profiles in laboratory boundary layers and channel and pipe flows, as well as in the near-neutral atmospheric surface layer. The von Karman constant is presumably a universal constant for all surface or wall layers. However, it is an empirical constant with a value of about 0.40; it has not been possible to determine it with an accuracy better than 5%.

Note that Eq. (10.5) also follows from the mixing length and eddy viscosity hypotheses, if one assumes that $l = kz$ and/or $K_m = kzu_*$ in the constant-flux surface layer. The integration of Eq. (10.5) with respect to z gives the well-known logarithmic velocity profile law,

$$U/u_* = (1/k) \ln(z/z_0) \tag{10.6}$$

in which z_0 has been introduced as a dimensional constant of integration and is commonly referred to as the roughness parameter or roughness length. The physical significance of z_0 and its relationship to surface roughness characteristics will be discussed later.

For a comparison of the power-law and logarithmic wind profiles, Eq. (10.6) can be written as

$$U/U_r = 1 + \ln(z/z_r)/\ln(z_r/z_0) \tag{10.7}$$

Figure 10.3 compares the plots of U/U_r versus z/z_r, according to Eqs.

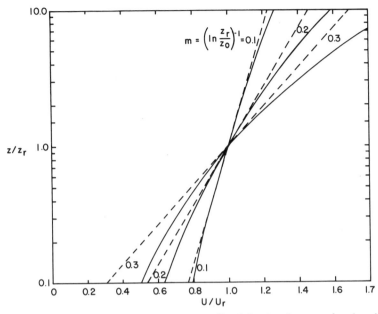

Fig. 10.3 Comparison of hypothetical wind profiles following the power law (——) [Eq. (10.2)] and the log law (---) [Eq. (10.7)] with their parameters related by Eq. (10.8).

(10.2) and (10.7), for different values of m and z_r/z_0, respectively. There is no exact correspondence between these two profile parameters, because the two profile shapes are different. An approximate relationship can be obtained by matching the velocity gradients, in addition to the velocities, at the reference height z_r:

$$m = \frac{d(\ln U)}{d(\ln z)} = \frac{z}{U}\frac{\partial U}{\partial z}\bigg|_{z=z_r} = \left(\ln \frac{z_r}{z_0}\right)^{-1} \tag{10.8}$$

Even with this type of matching, the two profiles deviate more and more as z deviates farther and farther from the matching height z_r. The theoretical relationship given by Eq. (10.8) between the power-law exponent m and the surface roughness parameter is also compared with observations in Fig. 10.2. For different data sets represented in that figure, z_r lies in the range between 10 and 100 m.

The logarithmic profile law has a sounder theoretical and physical basis than the power-law profile, especially within the neutral surface layer. The former implies linear variations of eddy viscosity and mixing length with height, irrespective of the surface roughness, while the power-law profile implies the unphysical, curvilinear behavior of K_m and l_m, even close to the surface.

Many observed wind profiles near the surface under near-neutral conditions have confirmed the validity of Eq. (10.6) up to heights of 20–200 m, depending on the PBL height. Figure 10.4 illustrates some of the observed wind profiles during the Wangara Experiment conducted over a low grass surface in southern Australia.

In the engineering fluid mechanics literature, Eq. (10.6) is referred to as the law of the wall, which has been found to hold in all kinds of pipe and channel flows, as well as in boundary layers. There, the distinction is made between aerodynamically smooth and rough surfaces. A surface is considered aerodynamically smooth if the small-scale surface protuberances or irregularities are sufficiently small to allow the formation of a laminar or viscous sublayer in which surface protuberances are com-

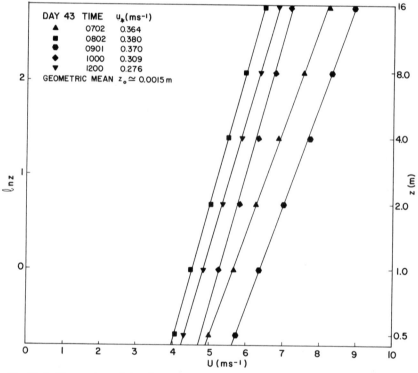

Fig. 10.4   Comparison of the observed wind profiles in the neutral surface layer of day 43 of the Wangara Experiment with the log law [Eq. (10.6)] (solid lines). [Data from Clarke *et al.* (1971).]

pletely submerged. If small-scale surface irregularities are large enough to prevent the formation of a viscous sublayer, the surface is considered aerodynamically rough. It is found from laboratory measurements that the thickness of the laminar sublayer in which the velocity profile is linear is about $5\nu/u_*$. For atmospheric flows the sublayer thickness would be of the order of 1 mm or less, while surface protuberances are generally larger than 1 mm. Therefore, almost all natural surfaces are aerodynamically rough; the only exceptions might be smooth ice, snow, and mud flats, as well as water surfaces under weak winds ($U_{10} < 3$ m sec^{-1}). The law of the wall (velocity-profile law) for an aerodynamically smooth surface is usually expressed in the form

$$\frac{U}{u_*} = \frac{1}{k} \ln \frac{u_* z}{\nu} + 5.1 \simeq \frac{1}{k} \ln \left(\frac{u_* z}{0.13\, \nu} \right) \tag{10.9}$$

in which the characteristic height scale of viscous sublayer (ν/u_*) is used in place of z_0 for a rough surface, and the numerical constant is evaluated from laboratory measurements in smooth flat-plate boundary layer and channel flows.

A comparison of Eqs. (10.6) and (10.9) shows that, if the former is used to represent wind profiles over a smooth surface, one should expect the roughness parameter to decrease with increasing u_*, according to the relation $z_0 \simeq 0.13\, \nu/u_*$. For a rigid, aerodynamically rough surface, however, z_0 is expected to remain constant, independent of u_*. Some surfaces are neither smooth nor completely rough, but fall into a transitional roughness regime. In nature, lake and ocean surfaces may fall into this transition between smooth and rough categories at moderate wind speeds ($2.5 < U_{10} < 7.5$ m sec^{-1}).

10.1.3 THE VELOCITY-DEFECT LAW

Above the surface layer, the Coriolis effects become significant and it is more appropriate to consider the deviations of actual velocity from the geostrophic velocity. For flat-plate boundary layer and channel flows, similarity consideration for the deviations of velocity from the ambient velocity leads to the so-called velocity-defect law

$$(U - U_\infty)/u_* = F(z/h) \tag{10.10}$$

The analogous geostrophic departure laws for the netrual barotropic PBL are

$$(U - U_g)/u_* = F_u(fz/u_*)$$
$$(V - V_g)/u_* = F_v(fz/u_*) \tag{10.11}$$

Here, the neutral PBL height is assumed to be proportional to u_*/f with an estimated coefficient of proportionality of about 0.3, so that $h \simeq 0.3u_*/f$.

A satisfactory confirmation of Eq. (10.11) and empirical determination of the similarity functions F_u and F_v cannot be made from atmospheric observations, because a strictly neutral, stationary, and barotropic PBL, which is not constrained by any inversion from above, is so rare in the atmosphere that relevant observations of the same do not exist. The averages of wind profiles measured under slightly unstable and slightly stable conditions appear to follow the above similarity scaling. Some laboratory simulations of the neutral Ekman layer in a rotating wind tunnel flow have also confirmed the same. Of course, there is plenty of experimental support for the validity of the velocity-defect law [Eq. (10.10)] in nonrotating flows.

10.2 SURFACE ROUGHNESS PARAMETERS

The aerodynamic roughness of a flat and uniform surface may be characterized by the average height (h_0) of the various roughness elements, their areal density, characteristic shapes, and dynamic response characteristics (e.g., flexibility and mobility). All the characteristics would be important, if one were interested in the complex flow field within the roughness or canopy layer. There is not much hope for a generalized and, at the same time, simple theoretical description of such a three-dimensional flow field in which turbulence dominates over the mean motion. In theoretical and experimental investigations of the fully developed surface layer, however, the surface roughness is characterized by only one or two roughness characteristics which can be empirically determined from wind-profile observations.

The roughness length parameter z_0 introduced in Eq. (10.6) is one such characteristic. In practice, z_0 is determined from the least-square fitting of Eq. (10.6) through the wind-profile data, or by graphically plotting $\ln z$ versus U and extrapolating the best-fitted straight line down to the level where $U = 0$, its intercept on the ordinate axis being $\ln z_0$. One should note however, that this is only a mathematical or graphical procedure for estimating z_0 and that Eq. (10.6) is not expected to describe the actual wind profile below the tops of roughness elements. An assumption implied in the derivation of (10.6) is that $z_0 \ll z$.

Empirical estimates of the roughness parameter for various natural surfaces can be ordered according to the type of terrain (see Fig. 10.5) or the average height of the roughness elements (see Fig. 10.6). Although z_0 varies over five orders of magnitude (from 10^{-5} m for smooth water sur-

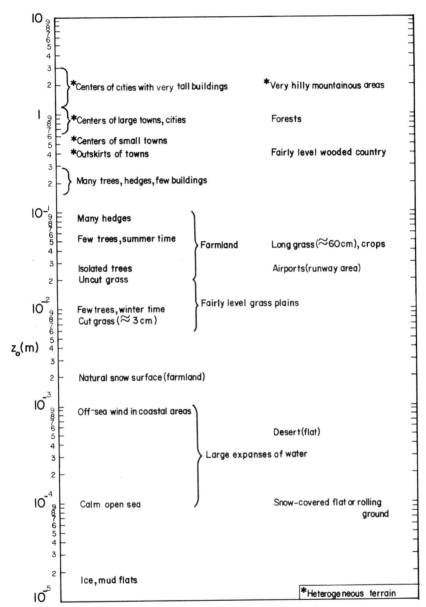

Fig. 10.5 Typical values, or range of values, of the surface roughness parameter for different types of terrain. [From tables by the Royal Aeronautical Society (1972).]

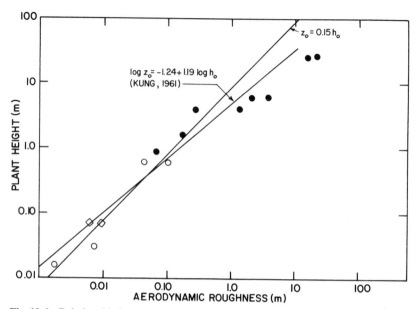

Fig. 10.6   Relationship between the aerodynamic roughness parameter and average vegetation height. ○, Deacon (1953); ●, Kung (1961); ◇, Chamberlain (1966). [From Plate (1971).]

faces to several meters for forests and urban areas), the ratio z_0/h_0 falls within a much narrower range (0.03–0.25) and increases gradually with increasing height of roughness elements. For uniform sand surfaces Prandtl and others have suggested a value of $z_0/h_0 \simeq 1/30$. An average value of $z_0/h_0 = 0.15$ has been determined for the various crops and grass lands; the same may also be used for many other natural surfaces (Plate, 1971). Figure 10.5 gives some typical values of z_0 for different types of surfaces and terrains. Figure 10.6 relates z_0 to h_0 for different types of crop canopies.

For very rough and undulating surfaces, the soil–air or water–air interface may not be the most appropriate reference datum for measuring heights in the surface layer. The air flow above the tops of roughness elements is dynamically influenced by the interface as well as by individual roughness elements. Therefore, it may be argued that the appropriate reference datum should lie somewhere between the actual ground level and the tops of roughness elements. In practice, the reference datum is determined empirically from wind-profile measurements in the surface

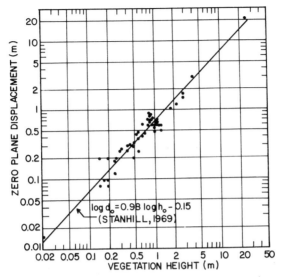

Fig. 10.7   Relationship between the zero-plane displacement and average vegetation height for different types of vegetation. [After Stanhill (1969).]

layer under near-neutral stability conditions. The modified logarithmic wind-profile law used for this purpose is

$$U/u_* = (1/k) \ln(z' - d_0)/z_0 \qquad (10.12)$$

in which d_0 is called zero-plane displacement or displacement length and z' is the height measured above the ground level.

For a plane surface, d_0 is expected to lie between zero and h_0, depending on the areal density of roughness elements. The displacement height may be expected to increase with increasing roughness density and approach a value close to h_0 for very dense canopies in which the flow within the canopy might become stagnant, or independent of the air flow above the canopy. These expectations are indeed borne out by empirical estimates of d_0 for vegetative canopies (see Fig. 10.7) which indicate that the appropriate datum for wind-profile measurements over most vegetative canopies is displaced above the ground level by 70–80% of the average height of vegetation. The zero-plane displacement in an urban boundary layer can similarly be expected to be a large fraction of the average building height. Figure 10.8 shows a verification of Eq. (10.12) against observed wind profiles over vegetative surfaces.

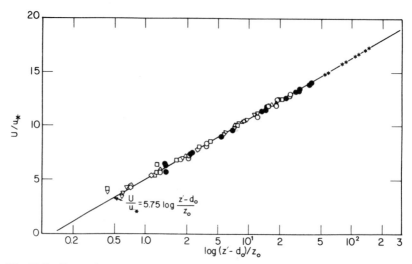

Fig. 10.8 Comparison of observed velocity profiles over crops with the log law [Eq. (10.12)]. ★, Snow; ●, low grass (flat terrain); ○, fallow; □, low grass; △, high grass; ▽, wheat; ◇, beets. [From Plate (1971).]

10.3   SURFACE STRESS AND DRAG COEFFICIENT

A direct measurement of the wind drag or stress on a relatively smooth surface can be made with an appropriate drag plate of a nominal diameter of 1–2 m and surface representative of the terrain around it. The larger the surface roughness, the more difficult it becomes for the drag plate to represent (simulate) such a roughness. Therefore, this method of drag measurement is limited to flat and uniform surfaces which are bare or have low vegetation. In most other cases, and also when drag-plate measurements cannot be made, the surface stress is determined indirectly from wind measurements in the surface layer. If mean winds are observed at a standard reference height z_r, the surface stress can be determined using a drag relation

$$\tau_0 = \rho C_D U_r^2 \qquad (10.13)$$

in which C_D is a dimensionless drag coefficient which depends on the surface roughness (more appropriately, on the ratio z_r/z_0) and atmospheric stability near the surface. In particular, for the near-neutral stability, the drag coefficient is given by

$$C_D = k^2/[\ln(z_r/z_0)]^2 \qquad (10.14)$$

which follows from Eqs. (10.6) and (10.13). Note that for a given z_r, C_D increases with increasing surface roughness. For a standard reference height of 10 m, values of C_D range from 1.2×10^{-3} for large lake and ocean surfaces at moderate wind speeds ($U_{10} < 6$ m sec^{-1}) to 7.5×10^{-3} for tall crops and moderately rough ($z_0 < 0.1$ m) surfaces. For very rough surfaces (e.g., forests and urban areas), however, a reference height of 10 m would not be adequate; it must be at least 1.5 times the height of roughness elements. In some applications the surface geostrophic wind (G_0) or the actual wind speed U_h at the top of the PBL is chosen as the reference velocity in Eq. (10.13).

Equation (10.13), with prescribed values of C_D, is often used for parameterizing the surface stress in large-scale atmospheric circulation models. Because specification of eddy viscosity also involves τ_0 or u_*, Eq. (10.13) can also be used in certain K models of the PBL. If detailed measurements of wind profile in the surface layer are available from a micrometeorological mast, u_* under near-neutral conditions can be determined from the least-square fitting of Eq. (10.6) to the profile data or from a single graphical plot of ln z versus U. Note that in the latter case, the slope of the best-fitted straight line must be equal to k/u_*.

10.4 TURBULENCE

Turbulence in a neutrally stratified boundary layer is entirely of mechanical origin and depends on the surface friction and vertical distribution of wind shear. From similarity considerations discussed in Section 10.3, the primary velocity scale is u_* and normalized standard deviations of velocity fluctuations (σ_u/u_*, σ_v/u_*, and σ_w/u_*) must be constants in the surface layer and some functions of the normalized height fz/u_* in the outer layer. Surface layer observations from different sites indicate that under near-neutral conditions $\sigma_u/u_* \simeq 2.5$, $\sigma_v/u_* \simeq 1.9$, and $\sigma_w/u_* \simeq 1.3$ (Panofsky and Dutton, 1984, Chap. 7). Observed values of turbulence intensity (σ_u/U) are shown in Fig. 10.9 as function of the roughness parameter. These are compared with the theoretical relation $\sigma_u/U = 1/\ln(z/z_0)$, which is based on $\sigma_u/u_* = 2.5$ and the logarithmic wind-profile law. The above theoretical relation seems to overestimate turbulence intensities over very rough surfaces. Counihan (1975) has suggested an alternative empirical relationship between σ_u/U and log z_0, which provides a better fit to the observed data in Fig. 10.9.

Measurements in the upper part of the neutral PBL are lacking, as pointed out earlier, but numerical model results indicate approximately exponential decrease of σ_u/u_*, etc., with height. For rough estimates of

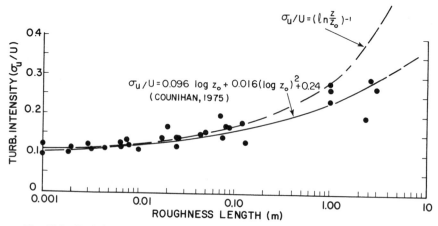

Fig. 10.9 Variation of turbulence intensity in the near-neutral surface layer with the roughness length. [Reprinted with permission from *Atmospheric Environment*, Vol. 9, J. C. Counihan, Adiabatic atmospheric boundary layers: A review and analysis of data from the period 1880–1972, Copyright (1975), Pergamon Journals Ltd.]

turbulent fluctuations, we suggest the following relations:

$$\sigma_u/u_* = 2.5 \exp|-3fz/u_*|$$
$$\sigma_v/u_* = 1.9 \exp|-3fz/u_*| \qquad (10.15)$$
$$\sigma_w/u_* = 1.3 \exp|-3fz/u_*|$$

The large-eddy length scale and Prandtl's mixing length are presumably proportional to each other. The latter is usually specified as

$$1/l_m = 1/kz + 1/l_0 \qquad (10.16)$$

which is an interpolation formula between the linear variation in the surface layer and the approach to a constant mixing length l_0 near the top of the PBL, which is assumed to be proportional to the PBL depth u_*/f or G/f.

10.5 APPLICATIONS

Semiempirical wind-profile laws and other relations discussed in this chapter may have the following practical applications:

• Determining the wind energy potential of a site
• Estimating wind loads on tall buildings and other structures
• Calculating dispersion of pollutants in very windy conditions

- Characterizing aerodynamic surface roughness
- Parameterizing the surface drag in large-scale atmospheric models, as well as in wave height and storm surge formulations.

PROBLEMS AND EXERCISES

1. Show that in the constant-stress surface layer, the power-law wind profile [Eq. (10.2)] implies a power-law eddy viscosity profile [Eq. (10.3)] and that the two exponents are related as $n = 1 - m$.

2. Derive the logarithmic wind-profile law on the basis of the eddy viscosity and mixing-lengths hypotheses, assuming $l_m = kz$, and $K_m = kzu_*$ in the constant-stress surface layer.

3. Using the logarithmic wind-profile law, plot the expected wind profiles (using a linear height scale) to a height of 100 m for the following combinations of the roughness parameter (z_0) and the mean wind speed at 100 m:

   (a) $z_0 = 0.01$ m; $U_{100} = 5, 10, 15$, and 20 m sec^{-1}

   (b) $U_{100} = 10$ m sec^{-1}; $z_0 = 10^{-3}, 10^{-2}, 10^{-1}$, and 1 m

   Also calculate for each profile the friction velocity u_* and the drag coefficient C_D for a reference height of 10 m, and show their values on the graphs.

4. The following mean velocity profiles were measured during the Wangara Experiment conducted over a uniform short-grass surface $(h_0 \approx 10$ cm$)$ under near-neutral stability conditions:

Height (m)	Mean wind speed (m sec^{-1})	
	Period 1	Period 2
0.5	7.82	4.91
1	8.66	5.44
2	9.54	6.06
4	10.33	6.64
8	11.22	7.17
16	12.01	7.71

   (a) Determine the roughness parameter and the surface stress for each of the above observation periods. Take $\rho = 1.25$ kg m^{-3}.

   (b) For both the periods estimate eddy viscosity, mixing length, and standard deviation of vertical velocity fluctuations at 20 m.

5. If the sea surface behaves like a smooth surface at low wind speeds

$(U_{10} < 2.5$ m sec$^{-1})$, like a completely rough surface at high wind speeds $(U_{10} > 7.5$ m sec$^{-1})$, and like a transitionally rough surface at speeds in between, express and plot the drag coefficient C_D as a function of U_{10} in the range $0.5 \leqslant U_{10}$ m sec$^{-1} \leqslant 50$, assuming the following:

(a) $z_0 = 0.13\nu/u_*$ in the aerodynamically smooth regime
(b) $z_0 = 0.02u_*^2/g$ in the aerodynamically rough regime
(c) $z_0 = 2 \times 10^{-4}$ m in the transitional regime

Chapter 11 | Momentum and Heat Exchanges with Homogeneous Surfaces

11.1 THE MONIN-OBUKHOV SIMILARITY THEORY

In the previous chapter, we showed that the atmospheric surface layer under neutral conditions is characterized by a logarithmic wind profile and nearly uniform (with respect to height) profiles of momentum flux and standard deviations of turbulent velocity fluctuations. It was also mentioned that the neutral stability condition is an exception rather than the rule in the lower atmosphere. More often, the turbulent exchange of heat between the surface and the atmosphere leads to thermal stratification of the surface layer and, to some extent, of the whole PBL, as shown in Chapter 5. It has been of considerable interest to micrometeorologists to find a suitable theoretical or semiempirical framework for a quantitative description of the mean and turbulence structure of the stratified surface layer. The Monin-Obukhov similarity theory has provided the most suitable and acceptable framework for organizing and presenting the micrometeorological data, as well as for extrapolating and predicting certain micrometeorological information where direct measurements of the same are not available.

11.1.1 THE SIMILARITY HYPOTHESIS

The basic similarity hypothesis first proposed by Monin and Obukhov (1954) is that in a horizontally homogeneous surface layer the mean flow and turbulent characteristics depend only on the four independent variables: the height above the surface z, the friction velocity u_*, the surface kinematic heat flux $H_0/\rho c_\mathrm{p}$, and the buoyancy variable g/T_0 (this appears in the expressions for buoyant acceleration and static stability given earlier). The simplifying assumptions implied in this similarity hypothesis are that the flow is horizontally homogeneous and quasistationary, the turbulent fluxes of momentum and heat are constant (independent of height), the molecular exchanges are insignificant in comparison with turbulent exchanges, the rotational effects can be ignored in the surface layer, and the influence of surface roughness, boundary layer height, and geostrophic winds is fully accounted for through u_*.

Because independent variables in the M–O similarity hypothesis involve three fundamental dimensions (length, time, and temperature), according to Buckingham's theorem, one can formulate only one independent dimensionless combination out of them. The combination traditionally chosen in the Monin–Obukhov similarity theory is the buoyancy parameter

$$\zeta = z/L$$

where

$$L = - u_*^3/[k(g/T_0)(H_0/\rho c_p)] \tag{11.1}$$

is an important buoyancy length scale, known as the Obukhov length after its originator. In his 1946 article in an obscure Russian journal (for the English translation, see Obukhov, 1946/1971), Obukhov introduced L as "the characteristic height (scale) of the sublayer of dynamic turbulence"; he also generalized the semiempirical theory of turbulence to the stratified atmospheric surface layer and used this approach to describe theoretically the mean wind and temperature profiles in the surface layer in terms of the fundamental stability parameter z/L.

One may wonder about the physical significance of the Obukhov length (L) and the possible range of its values. From the definition, it is clear that the values of L may range from $-\infty$ to ∞, the extreme values corresponding to the limits of the heat flux approaching zero from the positive (unstable) and the negative (stable) side, respectively. A more practical range of $|L|$, corresponding to fairly wide ranges of values of u_* and $|H_0|$ encountered in the atmosphere, is shown in Fig. 11.1. Here, we have assumed a typical value of $kg/T_0 = 0.013$ m sec^{-2} K^{-1}, which may not differ from the actual value of this by more than 15%. In magnitude $|L|$ represents the thickness of the layer of dynamic influence near the surface in which shear or friction effects are always important. The wind shear effects usually dominate and the buoyancy effects essentially remain insignificant in the lowest layer close to the surface ($z \ll |L|$). On the other hand, buoyancy effects may dominate over shear-generated turbulence for $z \gg |L|$. Thus, the ratio z/L is an important parameter measuring the relative importance of buoyancy versus shear effects in the stratified surface layer, similar to the Richardson number (Ri) introduced earlier. The negative sign in the definition of L is introduced so that the ratio z/L has

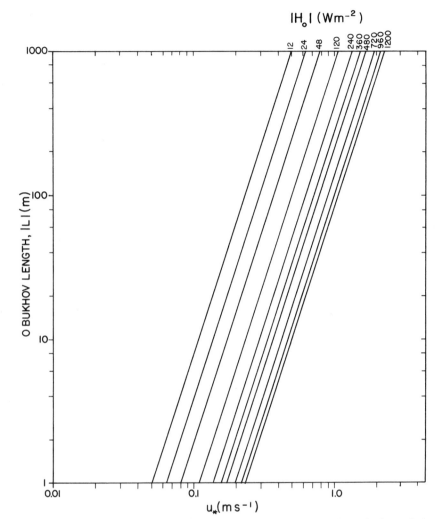

Fig. 11.1 Obukhov's buoyancy length as a function of the friction velocity and the surface heat flux.

the same sign as Ri. Later on, it will be shown that Ri and z/L are intimately related to each other, even though they have different distributions with respect to height in the surface layer (z/L obviously varies linearly with height, showing the increasing importance of buoyancy with height above the surface).

11.1.2 THE M–O SIMILARITY RELATIONS

The following characteristic scales of length, velocity, and temperature are used to form dimensionless groups in the Monin–Obukhov similarity theory:

z and L length scales
u_* velocity scale
$\theta_* \equiv -H_0/\rho c_p u_*$ temperature scale

The similarity prediction that follows from the M–O hypothesis is that any mean flow or average turbulence quantity in the surface layer, when normalized by an appropriate combination of the above-mentioned scales, must be a unique function of z/L only. Thus, a number of similarity relations can be written for the various quantities (dependent variables) of interest. For example, with the x axis oriented parallel to the surface stress or wind (the appropriate surface layer coordinate system), the dimensionless wind shear and potential temperature gradient are usually expressed as

$$(kz/u_*)(\partial U/\partial z) = \phi_m(\zeta)$$
$$(kz/\theta_*)(\partial \Theta/\partial z) = \phi_h(\zeta) \tag{11.2}$$

in which the von Karman constant k is introduced only for the sake of convenience, so that $\phi_m(0) = 1$, and $\phi_m(\zeta)$ and $\phi_h(\zeta)$ are the basic universal similarity functions which relate the constant fluxes

$$\tau = \tau_0 = \rho u_*^2$$
$$H = H_0 = -\rho c_p u_* \theta_* \tag{11.3}$$

to the mean gradients in the surface layer.

It is easy to show from the definition of Richardson number and Eqs. (11.1)–(11.2) that

$$\text{Ri} = \zeta\phi_h(\zeta)/\phi_m^2(\zeta) \tag{11.4}$$

which relates Ri to the basic stability parameter $\zeta = z/L$ of the M–O similarity theory. The inverse of Eq. (11.4), namely, $\zeta = f(\text{Ri})$ is often used to determine ζ and, hence, the Obukhov length L from easily measured gradients of velocity and temperature at one or more heights in the surface layer. Then, Eqs. (11.2) and (11.3) can be used to determine the fluxes of momentum and heat, knowing the empirical forms of the similarity functions $\phi_m(\zeta)$ and $\phi_h(\zeta)$ from carefully conducted experiments.

Other properties of flow which relate turbulent fluxes to the local mean gradients, such as the exchange coefficients of momentum and heat and the corresponding mixing lengths, can also be expressed in terms of $\phi_m(\zeta)$.

and $\phi_h(\zeta)$. For example, it is easy to show that

$$K_m/(kzu_*) = \phi_m^{-1}(\zeta)$$

$$K_h/(kzu_*) = \phi_h^{-1}(\zeta) \qquad (11.5)$$

$$K_h/K_m = \phi_m(\zeta)/\phi_h(\zeta)$$

which can be used to prescribe the eddy diffusivities or their ratio K_h/K_m in the stratified surface layer.

11.2 EMPIRICAL FORMS OF SIMILARITY FUNCTIONS

As mentioned earlier, a similarity theory based on a particular similarity hypothesis and dimensional analysis can only suggest plausible functional relationships between certain dimensionless parameters. It does not tell anything about the forms of those functions, which must be determined empirically from accurate observations during experiments, specifically designed for this purpose. Experimental verification is also required, of course, for the original similarity hypothesis or its consequences (e.g., predicted similarity relations).

Following the proposed similarity theory by Monin and Obukhov (1954), a number of micrometeorological experiments have been conducted at different locations (ideally, flat and homogeneous terrain with uniform and low roughness elements) and under fair-weather conditions (to satisfy the conditions of quasistationarity and horizontal homogeneity) for the expressed purpose of verifying the M–O similarity theory and for accurately determining the forms of the various similarity functions. Early experiments lacked direct and accurate measurements of turbulent fluxes, which could be estimated only after making some *a priori* assumptions about the flux-profile relations. The 1953 Great Plains Experiment at O'Neill, Nebraska, was one of the earliest concerted efforts in observing the PBL in which micrometeorological measurements were made by several different groups, using different instrumentation and techniques (Lettau and Davidson, 1957). But, here too, as in subsequent Australian and Russian experiments in the 1960s, the flux measurements suffered from severe instrument-response problems and were not very reliable. Still, these experiments could verify the general validity of M–O similarity theory and give the approximate forms of the similarity functions $\phi_m(\zeta)$ and $\phi_h(\zeta)$.

Perhaps the best micrometeorological experiment conducted so far, for the specific purpose of determining the M–O similarity functions, was the

1968 Kansas Field Program (Izumi, 1971). A 32-m tower located in the center of a 1-mile^2 field of wheat stubble ($h_0 \simeq 0.18$ m) was instrumented with fast-response cup anemometers, thermistors, resistance thermometers, and three-dimensional sonic anemometers at various levels. These were used to determine mean velocity and temperature gradients, as well as the momentum and heat fluxes using the eddy correlation method. In addition, two large drag plates installed nearby were used to measure the surface stress directly. Both heat and momentum fluxes were found to be constant with height, within the limits of experimental accuracy ($+20\%$, for fluxes). The flux-profile relations derived from the Kansas Experiment are discussed in detail by Businger *et al.* (1971).

The generally accepted forms of $\phi_m(\zeta)$ and $\phi_h(\zeta)$ on the basis of the 1968 Kansas Experiment and other experiments are

$$\phi_m = \begin{cases} (1 - \gamma_1\zeta)^{-1/4}, & \text{for} \quad \zeta < 0 \text{ (unstable)} \\ 1 + \beta\zeta, & \text{for} \quad \zeta \geqslant 0 \text{ (stable)} \end{cases} \qquad (11.6)$$

$$\phi_h = \begin{cases} \alpha(1 - \gamma_2\zeta)^{-1/2}, & \text{for} \quad \zeta < 0 \text{ (unstable)} \\ \alpha + \beta\zeta, & \text{for} \quad \zeta \geqslant 0 \text{ (stable)} \end{cases} \qquad (11.7)$$

There remain some differences, however, in the estimated values of the constants, α, β, γ_1, and γ_2 in the above expressions as obtained by different investigators. The main causes of these differences are the unavoidable measurement errors and deviations from the ideal conditions assumed in the theory. The best estimated values from the Kansas Experiment are (Businger *et al.*, 1971)

$$\alpha = 0.74; \qquad \beta = 4.7; \qquad \gamma_1 = 15; \qquad \gamma_2 = 9 \qquad (11.8)$$

The corresponding formulas [Eqs. (11.6) and (11.7)] for ϕ_m and ϕ_h with the above empirical values of constants are shown in Figures 11.2 and 11.3 together with the observed Kansas data.

In view of the above-mentioned uncertainties in measurements and in the determination of empirical similarity functions or constants, the following simpler flux-profile relations can be recommended for most practical applications in which great precision is not warranted:

$$\phi_h = \phi_m^2 = (1 - 15\zeta)^{-1/2}, \qquad \text{for} \quad \zeta < 0$$

$$\phi_h = \phi_m = 1 + 5\zeta, \qquad \text{for} \quad \zeta \geqslant 0 \qquad (11.9)$$

These deviate from the Kansas relations [Eqs. (11.6)–(11.8)] only slightly, but have the advantage of relating ζ to Ri more simply and explicitly as

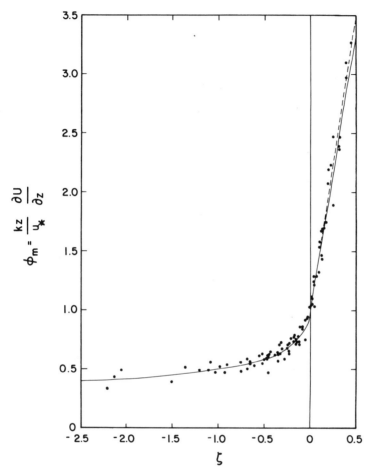

Fig. 11.2 Dimensionless wind shear as a function of the M–O stability parameter ζ. ——, Eqs. (11.6)–(11.8); ---, Eq. (11.9); ·, Kansas data. [Data from Izumi (1971).]

$$\zeta = Ri, \qquad \text{for} \qquad Ri < 0$$

$$\zeta = \frac{Ri}{1 - 5Ri}, \qquad \text{for} \qquad 0 \leqslant Ri \leqslant 0.2$$

(11.10)

These are verified against the Kansas data in Fig. 11.4.

After substituting from Eqs. (11.9) and (11.10) into Eq. (11.5), the eddy diffusivities of heat and momentum can also be expressed as functions of

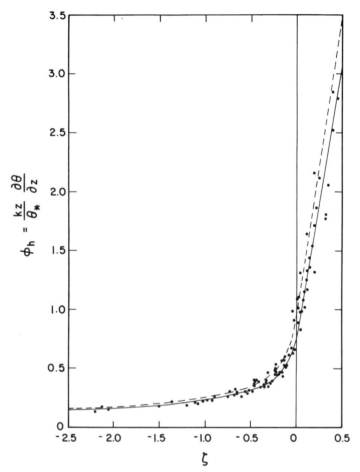

Fig. 11.3 Dimensionless potential temperature gradient as a function of the M–O stability parameter ζ. ——, Eqs. (11.7) and (11.8); ---, Eq. (11.9); ·, Kansas data. [Data from Izumi (1971).]

the Richardson number

$$\frac{K_m}{kzu_*} = \begin{cases} (1 - 15\text{Ri})^{1/4}, & \text{for} \quad \text{Ri} < 0 \\ 1 - 5\text{Ri}, & \text{for} \quad 0 \leqslant \text{Ri} \leqslant 0.2 \end{cases}$$

$$\frac{K_h}{kzu_*} = \begin{cases} (1 - 15\text{Ri})^{1/2}, & \text{for} \quad \text{Ri} < 0 \\ 1 - 5\text{Ri}, & \text{for} \quad 0 \leqslant \text{Ri} \leqslant 0.2 \end{cases} \qquad (11.11)$$

These are represented in Fig. 11.5. Note that the eddy diffusivities of heat and momentum are equal under stably stratified conditions; they decrease

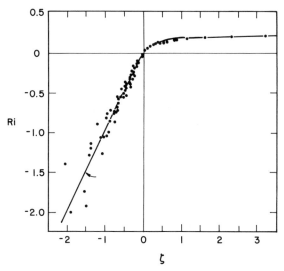

Fig. 11.4   The Richardson number as a function of the M–O stability parameter ζ. Arrow, Eq. (11.10). [After Businger *et al.* (1971).]

rapidly with increasing stability and vanish as the Richardson number approaches its critical value of $Ri_c = 1/\beta \simeq 0.2$.

There is a question about the validity of the above-mentioned empirically estimated similarity functions under extremely unstable (approach-

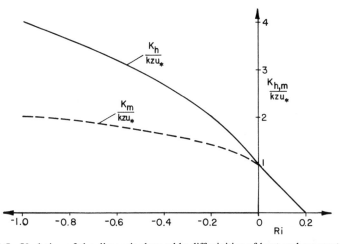

Fig. 11.5   Variation of the dimensionless eddy diffusivities of heat and momentum with Richardson number according to Eq. (11.11).

ing free convection) and extremely stable (approaching critical Ri) conditions. Micrometeorological data used in the determination of the above M–O similarity functions have generally been limited to the moderate stability range of $-5 < \zeta < 2$, and, as such, are strictly valid in this range. As a matter of fact, in conditions approaching free convection ($-\zeta \gg 1$, or $\zeta \rightarrow -\infty$), the above expressions for ϕ_h and K_h are not quite consistent with the local free convection similarity relations for $\partial\Theta/\partial z$ and K_h derived in Chapter 9. In order for them to be consistent one must have $\phi_h \sim (-\zeta)^{-1/3}$ for $-\zeta \gg 1$.

On the other side of strong stability conditions, there is some evidence of a linear velocity profile in the lower part of the PBL, including the surface layer, which appears to be consistent with the M–O relation for $\phi_m(\zeta)$ for $\zeta \gg 1$. There is also other experimental evidence suggesting the limiting value of $\zeta \simeq 1$ for the validity of the M–O relation $\phi_m = 1 + \beta\zeta$. In any case, strong stability conditions usually occur at nighttime under clear skies and weak winds. Because the temperature profile under such conditions is strongly influenced by longwave atmospheric radiation, which is ignored in the M–O similarity theory, the temperature field may not follow the M–O similarity.

11.3   WIND AND TEMPERATURE PROFILES

Integration of Eqs. (11.2) with respect to height yields the velocity and potential temperature profiles in the form

$$U/u_* = (1/k)[\ln(z/z_0) - \psi_m(z/L)]$$

$$(\Theta - \Theta_0)/\theta_* = (1/k)\,[\ln(z/z_0) - \psi_h(z/L)] \tag{11.12}$$

in which Θ_0 is the extrapolated temperature at $z = z_0$ and ψ_m and ψ_h are different similarity functions related to ϕ_m and ϕ_h, respectively, as

$$\psi_m\left(\frac{z}{L}\right) = \int_{z_0/L}^{z/L} [1 - \phi_m(\zeta)]\,\frac{d\zeta}{\zeta}$$

$$\psi_h\left(\frac{z}{L}\right) = \int_{z_0/L}^{z/L} [1 - \phi_h(\zeta)]\,\frac{d\zeta}{\zeta} \tag{11.13}$$

For smooth and moderately rough surfaces, z_0/L is usually quite small and the integrands in Eq. (11.13) are well behaved at small ζ, so that the lower limit of integration can essentially be replaced by zero. With this approximation, the ψ functions can be determined for any appropriate forms of ϕ

functions. For example, corresponding to Eq. (11.9), we obtain

$$\psi_m = \psi_h = -5\frac{z}{L}, \qquad\qquad \text{for } \frac{z}{L} \geq 0$$

$$\psi_m = \ln\left[\left(\frac{1+x^2}{2}\right)\left(\frac{1+x}{2}\right)^2\right] - 2\tan^{-1} x + \frac{\pi}{2}, \qquad \text{for } \frac{z}{L} < 0 \qquad (11.14)$$

$$\psi_h = 2\ln\left(\frac{1+x^2}{2}\right), \qquad\qquad \text{for } \frac{z}{L} < 0$$

where $x = (1 - 15z/L)^{1/4}$.

Note that the deviations in profiles from the log law increase with increasing magnitude of z/L. Under stable conditions the profiles are log linear and tend to become linear for large values of z/L.

Under unstable conditions ($\zeta < 0$), on the other hand, ψ_m and ψ_h are positive, so that the deviations from the log law are of opposite sign (negative). Consequently, the velocity and temperature profiles in the surface layer are expected to become more and more curvilinear as instability increases. The observed winds and temperatures are usually plotted against $\log z$ or $\ln z$, instead of a linear height scale, in order to reduce the profile curvature and to clearly show their deviations from the log law due to buoyancy effects in the surface layer.

11.4 DRAG AND HEAT TRANSFER COEFFICIENTS

The surface stress and sensible heat flux are usually expressed or parameterized in terms of dimensionless drag and heat transfer coefficients as

$$\tau_0 = \rho C_D U^2$$

$$H_0 = -\rho c_p C_H U(\Theta - \Theta_0) \qquad (11.15)$$

in which C_D and C_H are the drag and heat transfer coefficients, respectively, and U and Θ are the mean wind speed and potential temperature at the measurement height z. When z falls within the surface layer, it is easy to show from Eqs. (11.12) and (11.15) that

$$C_D = k^2[\ln(z/z_0) - \psi_m(z/L)]^{-2}$$

$$C_H = k^2[\ln(z/z_0) - \psi_m(z/L)]^{-1}[\ln(z/z_0) - \psi_h(z/L)]^{-1} \qquad (11.16)$$

Thus, according to the Monin–Obukhov similarity theory, the drag and heat transfer coefficients are some universal functions of z/z_0 and z/L. Because the latter parameter involves the difficult-to-determine Obukhov length L, it should be useful and desirable to relate z/L to a more easily estimated parameter from the known variables U, Θ, Θ_0, and z. One such parameter is the bulk Richardson number

$$\mathrm{Ri_B} = (g/T_0)[(\Theta - \Theta_0)z/U^2] \tag{11.17}$$

Which can be related to z/L by substituting into Eq. (11.17) from the M–O profile relations given by Eq. (11.12). It is easy to show that

$$\mathrm{Ri_B} = \frac{z}{L}\left[\ln\frac{z}{z_0} - \psi_h\left(\frac{z}{L}\right)\right]\left[\ln\frac{z}{z_0} - \psi_m\left(\frac{z}{L}\right)\right]^{-2} \tag{11.18}$$

which is of the form $\mathrm{Ri_B} = F(z/z_0, z/L)$. The inverse of this function can be used to determine z/L for given values of z/z_0 and $\mathrm{Ri_B}$. For this a graphical representation of Eq. (11.18) would be useful. For stable conditions, the inverse relationship is exactly given by

$$\frac{z}{L} = \frac{\mathrm{Ri_B}}{1 - 5\mathrm{Ri_B}}\ln\frac{z}{z_0}, \quad \text{for } 0 \leqslant \mathrm{Ri_B} < 0.2 \tag{11.19}$$

Knowing the general forms of ψ functions one can infer from Eq. (11.16) that both the drag and heat transfer coefficients increase with

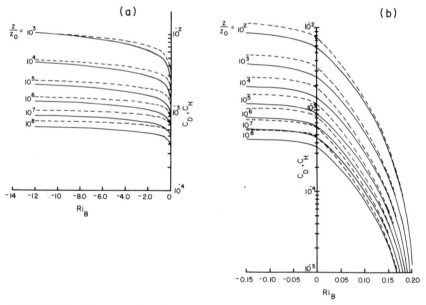

Fig. 11.6   Variation of surface drag and heat transfer coefficients with surface roughness and the bulk Richardson number for (a) unstable conditions and (b) near-neutral and stable conditions. ———, C_D; ---, C_H. [After Arya (1977).]

increasing surface roughness (decreasing z/z_0) and decrease with increasing stability (see Fig. 11.6). Their typical values over comparatively smooth water surfaces range from 1.0×10^{-3} to 2×10^{-3}. For land surfaces, however, the range of C_D and C_H values is considerably wider (say, 0 to 0.01). These are only weakly dependent on the choice of observation level in the surface layer; the most frequently recommended observation height for marine meteorological observations is 10 m.

11.5 METHODS OF DETERMINING MOMENTUM AND HEAT FLUXES

Determination of turbulent exchanges taking place between the earth and the atmosphere near their interface (surface) is of primary concern in micrometeorology. A number of methods have been devised with varying degrees of sophistication, some of which will be described here only briefly. The emphasis here is on principles and techniques, rather than on the details of instrumentation and measurements.

11.5.1 SURFACE-DRAG MEASUREMENTS

The only direct method of measuring shearing stress on a small sample of the surface is through the use of a carefully installed drag plate. The drag measured on the sample surface (plate) area must be representative of the whole area under consideration. This presumes a reasonably uniform ground cover with small roughness elements, which remain relatively undisturbed by the installation of the drag plate. The original roughness characteristics of the surface are retained or duplicated on the top of the drag plate. The drag plate has a small annular ring around to isolate it from the surrounding area for drag force measurements using strain gauges and electromechanical transducers. The installation and successful operation of a drag plate requires considerable care, skill, and experience. For this reason, such measurements have been made only in the context of certain micrometeorological experiments or research expeditions (e.g., the 1968 Kansas Field Program).

11.5.2 ENERGY BALANCE METHOD

No method exists for directly measuring the surface heat flux. A thin heat flux plate of known conductivity, with an embedded thermopile to measure temperature gradient across the plate, is often used to measure heat flux through the subsurface medium (e.g., soil, ice, and snow). But, to avoid radiative and convective effects the plate must be buried at least

10 mm below the surface. One can determine the ground heat flux H_G after applying an appropriate correction to the heat flux plate measurements. The sensible heat flux at the surface to or from air can be estimated indirectly, using the surface energy balance or other appropriate energy budget equation discussed in Chapter 2, provided other components of the energy balance are measured or can otherwise be estimated.

11.5.3 EDDY CORRELATION METHOD

The most reliable and direct measurements of turbulent exchanges of momentum and heat in the atmosphere are usually made with sophisticated fast-response turbulence instrumentation. If all fluctuations of velocity and temperature that contribute to the desired momentum and heat fluxes are faithfully sensed and recorded, one can determine their covariances simply by averaging the products of the appropriate fluctuations over any desired averaging time. In particular, the vertical fluxes of momentum and heat over a homogeneous surface are given by

$$\tau = -\rho\overline{uw}$$
$$H = \rho c_p \overline{\theta w}$$

(11.20)

with the x axis oriented along the mean wind. Because the above fluxes are expected to remain constant, independent of height, in a horizontally homogeneous surface layer, eddy correlation measurements also provide a means of determining the surface fluxes ($\tau_0 \simeq -\rho\overline{uw}$ and $H_0 \simeq \rho c_p \overline{\theta w}$).

Although the eddy correlation method of determining fluxes is simple, in practice, it requires expensive research-grade instrumentation, such as sonic, laser, or hot-wire anemometers and fine-resistance thermometers, as well as rapid (sampling rates of $10–100 \text{ sec}^{-1}$) data acquisition systems. The closer to the surface the turbulence measurements are made, the more severe become the instrument response problems. At heights of larger than 10 m or so, light cup, vane, and propeller anemometers may also be adequate for measuring variances and fluxes. Larger averaging times may be required, however, with increasing height of measurement, because the characteristic size of large eddies usually increases with height in the PBL. The requirements of instrument leveling, orientation, calibration, and maintenance are also quite severe for accurate eddy correlation measurements.

The above-mentioned instrumental requirements have kept the eddy correlation method from being widely used, except in special research expeditions. But, continued advances in instruments and data processing might make it more practical in the near future. The method has the advantage of measuring turbulent exchanges directly, without too many restrictive assumptions about the nature of the surface (such as uniform, flat, and homogeneous) or of the atmosphere. It is the only method available for measuring turbulent fluxes inside plant canopies, or in the wakes of hills and buildings.

11.5.4 BULK TRANSFER METHOD

Indirect methods of estimating fluxes from more easily measured mean winds and temperatures in the surface layer or the whole PBL are based on the appropriate flux-profile relations. The simplest and the most widely used method is the bulk aerodynamic approach, which is based on the bulk transfer formulas [Eq. (11.15)]. This method can be used when measurements or computations (e.g., in a numerical model) of mean velocity and temperature are available only at one level, in conjunction with the desired surface properties (e.g., the surface roughness and temperature). If the observation height is low enough for it to fall in the surface layer, the appropriate drag and heat transfer coefficients in Eq. (11.15) can be parameterized on the basis of the Monin–Obukhov similarity relations [Eq. (11.16)], as described in Section 11.4. This will indeed be the case for most routine surface meteorological observations over land and ocean areas. On the other hand, if the observation level is near the top of the PBL, or the reference wind is geostrophic, C_D and C_H may be parameterized on the basis of the PBL similarity theory (see, e.g., Deardorff, 1972; Arya, 1984). In practice C_D and C_H over ocean surfaces are usually prescribed some constant average values (e.g., $C_D \simeq C_H \simeq 1.5 \times 10^{-3}$) derived from major marine meteorological experiments. Over land areas, C_D and C_H are found to vary over a much greater range, due to the effects of surface roughness and stability, but in many applications constant values are still prescribed for the sake of simplicity. It is not difficult, however, to incorporate in such parameterizations the stability-dependent correction factors, such as C_D/C_{DN} and C_H/C_{HN} as functions of Ri_B (see Deardorff, 1968), where C_{DN} and C_{HN} are the prescribed coefficients for neutral stability.

11.5.5 GRADIENT OR AERODYNAMIC METHOD

In order to use the bulk transfer method described above, one needs to know the surface roughness and the surface temperature. Such information is not always easy to come by. In fact, for very rough and uneven surfaces, the surface itself is not well defined and its temperature cannot be measured directly. This difficulty can be avoided by making measurements at two or more heights in the surface layer. Here, we describe a simple gradient or aerodynamic method of determining fluxes from measurements of mean differences or gradients of velocity and temperature between any two heights z_1 and z_2 within the surface layer, but well above the tops of roughness elements.

Let $\Delta U = U_2 - U_1$ and $\Delta \Theta = \Theta_2 - \Theta_1$ be the difference in mean velocities and potential temperatures (note that $\Delta \Theta = \Delta T + \Gamma \Delta z$) across the height interval $\Delta z = z_2 - z_1$. One can determine the gradient Richardson number at the geometric mean height $z_m = (z_1 z_2)^{1/2}$ by writing

$$\text{Ri}(z_m) = \frac{g}{T_0} \frac{\partial \Theta/\partial z}{(\partial U/\partial z)^2}\bigg|_{z_m} = \frac{gz}{T_0} \frac{\partial \Theta/\partial \ln z}{(\partial U/\partial \ln z)^2}\bigg|_{z_m} \tag{11.21}$$

The finite-difference approximations used for the gradients in the surface layer are

$$\frac{\partial \Theta}{\partial \ln z} \simeq \frac{\Delta \Theta}{\Delta \ln z} = \frac{\Delta \Theta}{\ln(z_2/z_1)}$$

$$\frac{\partial U}{\partial \ln z} \simeq \frac{\Delta U}{\Delta \ln z} = \frac{\Delta U}{\ln(z_2/z_1)} \tag{11.22}$$

Here, $\ln z$ is used instead of z, because the profiles are more nearly linear with respect to the former. Then, the Richardson number is approximately given as

$$\text{Ri}(z_m) \simeq \frac{g}{T_0} z_m \left(\ln \frac{z_2}{z_1}\right) \frac{\Delta \Theta}{(\Delta U)^2} \tag{11.23}$$

The corresponding value of the Monin–Obukhov stability parameter $\zeta_m = z_m/L$ can be determined from the relations given by Eq. (11.10)

$$z_m/L = \text{Ri}(z_m), \qquad \qquad \text{for} \qquad \text{Ri} < 0$$

$$z_m/L = \text{Ri}(z_m)/[1 - 5\text{Ri}(z_m)], \qquad \text{for} \quad 0 \leq \text{Ri} < 0.2 \tag{11.24}$$

Then $\phi_m(\zeta_m)$ and $\phi_h(\zeta_m)$ are determined from Eq. (11.9), and u_* and θ_* from Eq. (11.2), again using the finite-difference approximations [Eq. (11.22)]. Thus, the surface shear stress and heat flux are given by

$$\tau_0 = \rho u_*^2 = \rho \left[\frac{k\Delta U}{\phi_m(\zeta_m)\ln(z_2/z_1)}\right]^2 \tag{11.25}$$

$$H_0 = -\rho c_p u_* \theta_* = -\rho c_p \left[\frac{k^2 \Delta U \Delta \Theta}{\phi_m(\zeta_m)\phi_h(\zeta_m)\left(\ln \frac{z_2}{z_1}\right)^2}\right]$$

Alternatively, one can determine the exchange coefficients of heat and momentum by using Eq. (11.11), and then calculate the fluxes using the gradient-transfer relations

$$\tau_0 = \rho K_m (\partial U / \partial z)$$

$$H_0 = -\rho c_p K_h (\partial \Theta / \partial z)$$

(11.26)

The finite-difference approximations [Eq. (11.22)] may not be good enough when the ratio z_2/z_1 becomes very large. On the other hand, if the two height levels are too close to each other, the differences in velocities and temperatures at the two levels may not be well resolved. A good compromise is to specify the measurement heights such that $z_2/z_1 = 2$ to 4. Still, there must be expected some errors in the measurements of ΔU and $\Delta \Theta$, as well as in the estimation of $\partial U/\partial z$, $\partial \Theta/\partial z$, and $\mathrm{Ri}(z_m)$. These, in conjunction with uncertainties in the M–O flux-profile relations, may lead to errors of more than 30% in the estimates of surface fluxes. Due to the above-mentioned errors and other reasons, the estimated value of $\mathrm{Ri}(z_m)$ in stably stratified conditions may even turn out to be greater than the critical value of 0.2. The gradient method becomes invalid and should not be used in such cases, unless appropriate values of eddy diffusivities in such strongly stratified conditions can be prescribed for use in Eq. (11.26), without recourse to the M–O similarity theory.

11.5.6 PROFILE METHOD

In order to minimize errors in the estimated fluxes, it is highly desirable to make measurements of mean velocity and temperature at several (more than two) levels within the surface layer. A good procedure for determining fluxes from such profile measurements is to fit the appropriate flux-profile relations to the observations, using the least-square technique. In this way, the effect of random experimental errors on flux estimates can be minimized. A simple graphical procedure can also be used for the same purpose, which is described here in the following.

First, calculate Ri from measurements of U and Θ at each pair of consecutive levels and obtain the best estimate of the Obukhov length L by fitting a straight line through the data points of z_m versus Ri plot for unstable conditions, or z_m versus $\mathrm{Ri}/(1 - 5\mathrm{Ri})$ plot for stable conditions. Note that in either case, according to Eq. (11.24), the slope of the best-fitted line will be L. The second step is to plot U versus $\ln z - \psi_m(z/L)$ and Θ versus $\ln z - \psi_h(z/L)$ and draw best-fitted straight lines through these data points (see Figs. 11.7 and 11.8). Note that the slopes of these lines, according to Eq. (11.12), must be u_*/k and θ_*/k, respectively, which readily determine the surface fluxes. The additional information on intercepts can be used to determine z_0 and Θ_0 if desired. This follows from

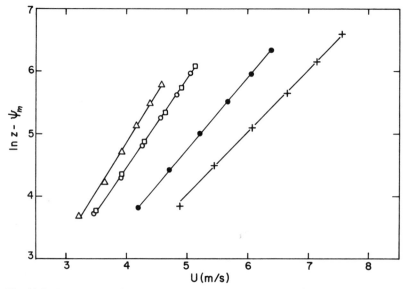

Fig. 11.7   Least-square fitting of the M–O flux profile relation (modified log law) to the observed mean velocity profiles at Kerang, Australia. Run: +, K-1; ●, K-2; □, K-3; ○, K-4; △, K-5. [After Paulson (1967).]

writing Eq. (11.12) in the form

$$U = \frac{u_*}{k}\left[\ln z - \psi_m\left(\frac{z}{L}\right)\right] - \frac{u_*}{k}\ln z_0$$

$$\Theta = \frac{\theta_*}{k}\left[\ln z - \psi_h\left(\frac{z}{L}\right)\right] + \left(\Theta_0 - \frac{\theta_*}{k}\ln z_0\right)$$

(11.27)

in which the last terms are independent of height.

The above procedure is an extension of the more familiar graphical method of determining the friction velocity or the surface stress from the observed wind profile in a neutral surface layer (see Chapter 10). The stability correction is made here to obtain a modified height coordinate in which the wind and potential temperature profiles are expected to be linear.

11.5.7   GEOSTROPHIC DEPARTURE METHOD

The equations of mean motion (see, e.g., Chapter 9), with certain assumptions, can be used to determine the surface stress, as well as the profiles of turbulent momentum fluxes in the PBL. Here, we demonstrate this method for a horizontally homogeneous and quasistationary PBL for

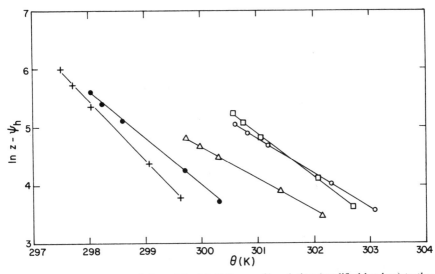

Fig. 11.8 Least-square fitting of the M–O flux profile relation (modified log law) to the observed mean potential temperature profiles at Kerang, Australia. Run: +, K-1; ●, K-2; □, K-3; ○, K-4; △, K-5. [After Paulson (1967).]

which the mean flow equations reduce to Eq. (9.9). Their integration with respect to height from some level z to the top of the PBL, where the turbulent fluxes may be expected to vanish, gives momentum fluxes as functions of the height z

$$\overline{uw} = -f \int_z^h (V - V_g)\, dz$$

$$\overline{vw} = f \int_z^h (U - U_g)\, dz$$

$$(11.28)$$

In particular, the surface stress is given by

$$\tau_0 = \rho f \int_0^h (V - V_g)\, dz \qquad (11.29)$$

where the x axis is taken parallel to the surface wind or stress [this implies that $\overline{vw} = 0$ at $z = 0$, so that from Eq. (11.28) $\int_0^h (U - U_g)\, dz = 0$ provides a check]. The above relations are, of course, restricted to the steady-state flow with no advection, but can easily be modified to include acceleration terms.

Note that Eqs. (11.28) and (11.29) express the local momentum fluxes and the surface stress in terms of the integrated departures of the actual wind components from the geostrophic wind components. Hence, the name "geostrophic departure method" is given to this simple technique of determining turbulent fluxes from more easily measured wind profiles (e.g., from pibals, rawinsondes, or remote-sensing wind profilers). The presence of accelerations, and uncertainties in the determination of actual mean winds, geostrophic winds, and the PBL height, frequently lead to deviations from the ideal case and, hence, to large errors in flux determinations. Because the top of the PBL where turbulent momentum fluxes are expected to vanish may not be easy to identify from mean wind profiles alone, the upper limit of integrals in Eqs. (11.28) and (11.29) is sometimes replaced by the heights z_u and z_v, where the U and V profiles have a maximum or minimum, provided such maxima or minima exist. This is based on the assumption, implied by the gradient-transport hypothesis, that $\overline{uw}$ and $\overline{vw}$ must vanish at the levels where $\partial U/\partial z$ and $\partial V/\partial z$ become zero, respectively.

11.5.8 THERMODYNAMIC ENERGY EQUATION METHOD

A similar method for determining the sensible heat flux at the surface and its vertical distribution (profile) in the PBL is based on the thermodynamic energy equation which, in the absence of temperatures advection, reduces to

$$\frac{\partial \Theta}{\partial t} = \frac{1}{\rho c_p} \frac{\partial R_N}{\partial z} - \frac{\partial \overline{w\theta}}{\partial z} \tag{11.30}$$

Note that the time-tendency (warming or cooling rate) term is retained here, because it is often found to be significant even when the flow field may be considered quasistationary. It is a manifestation of the diurnal heating and cooling cycle which is responsible for the important stability and buoyancy effects in the PBL. According to Eq. (11.30), the rate of warming or cooling essentially balances the convergence or divergence of radiative and sensible heat fluxes. The radiative flux divergence is usually ignored in the daytime unstable or convective PBL, especially in the absence of fog and clouds within the layer. It becomes more significant in the stably stratified nocturnal boundary layer.

In simpler situations, where the radiative flux divergence can be ignored, the integration of Eq. (11.30) with respect to height yields

$$\overline{w\theta} = \int_z^h \frac{\partial \Theta}{\partial t}\, dz$$

$$H_0 = \rho c_p \int_0^h \frac{\partial \Theta}{\partial t}\, dz \tag{11.31}$$

in which we have assumed that the sensible heat flux vanishes at the top of the PBL, i.e., at $z = h$, $\overline{w\theta} = 0$. Thus, temperature soundings in the PBL at close intervals (say 1–3 hr) may be used with Eq. (11.31) to determine H_0 and the $\overline{w\theta}$ profile. The above relations must be restricted to the case of no horizontal and vertical temperature advections, but can easily be modified to include advection terms.

Both the geostrophic departure and thermodynamic energy methods are quite useful and reliable in that these are based on the fundamental conservation equations and measurements of mean wind and temperature profiles, without any restrictive assumptions, as implied in the empirical flux-profile relations. The simplifying assumptions of steady state and no advection can be relaxed whenever they cannot be justified and advection/acceleration terms can be evaluated from mean profile measurements.

11.5.9 VARIANCE METHOD

In a horizontally homogeneous and quasistationary PBL the turbulent fluxes or covariances are intimately related to the variances or standard deviations of velocity and temperature fluctuations, through the various similarity relations for turbulence structure. Since variances can be measured more easily and accurately than covariances or turbulent fluxes, the latter may be determined indirectly by measuring the former and using the appropriate similarity relations. Still, fast-response instruments are required, which is a disadvantage in comparison with other indirect methods requiring measurements of mean winds and temperatures. However, the fact that observations at a single level may yield momentum and heat fluxes is an added attraction of the variance method.

The method is best suited for determining the surface fluxes from measurements of σ_w and σ_θ at an appropriate level (say, 10 m) in the fully developed surface layer, because the similarity relations for the same are simpler and well established for a wide range of surface types and stability conditions. For example, in the unstable surface layer both σ_w and σ_θ follow the local free convection similarity relations [Eq. (9.29)] (see Fig. 11.9), either of which can be used to determine the surface heat flux. In the stably stratified surface layer, on the other hand, the local similarity scaling of turbulence structure implies that the ratios σ_w/u_* and σ_θ/θ_* must be constants, independent of height, where $u_* = (-\overline{uw})^{1/2}$ and $\theta_* = -\overline{w\theta}/u_*$. This can be used to estimate the local fluxes from measurements of σ_w and σ_θ. In slightly unstable to slightly stable conditions, the Monin–Obukhov similarity theory predicts σ_w/u_* and σ_θ/θ_* to be some unique functions of z/L, which have been evaluated on the basis of micromete-

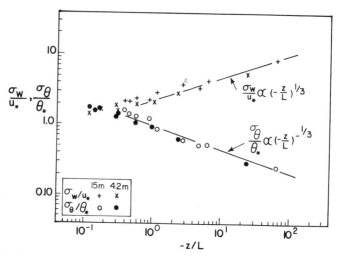

Fig. 11.9 Normalized standard deviations of vertical velocity and temperature fluctuations as functions of $-z/L$, compared with local free-convection similarity relations. [From Businger (1973); after Monji (1972).]

orological data from homogeneous sites. In particular, the empirical formula

$$\frac{\sigma_w}{u_*} = 1.25[1 - 3(z/L)]^{1/3} \tag{11.32}$$

obtained from a compilation of many aircraft and tower data taken over land and sea can be recommended for unstable conditions (see also Panofsky and Dutton, 1984, Chap. 7).

The surface heat flux in a convective boundary layer (CBL) can also be estimated from measurements of σ_u or σ_v at any convenient height in the lower part of the CBL where the empirical similarity relations $\sigma_u \simeq \sigma_v \simeq 0.6W_*$ are found to be valid (Caughey and Palmer, 1979). Since, the convective velocity scale W_* is related to the mixed-layer height and the surface heat flux, the latter can be estimated if the mixed-layer height is measured independently (e.g., from temperature sounding).

In some applications, the above-mentioned similarity relations between the turbulent fluxes and variances are used to determine the latter from indirect estimates of the former. For example, the horizontal and vertical spread of a chimney plume or a pollutant cloud is intimately related to velocity variances, which need to be indirectly estimated in the absence of turbulence measurements.

11.6   APPLICATIONS

Theory and observations of momentum and heat exchanges between the atmosphere and the earth's surface, presented in this chapter, may have the following practical applications:

- Systematic ordering of mean wind, temperature, and turbulence data with stability in the surface layer
- Quantitative description of the surface layer wind and temperature profiles and their relationship to the surface fluxes of momentum and heat
- Relating different stability parameters
- Specifying eddy diffusivities and bulk transfer coefficients as functions of stability in the parameterizations of fluxes
- Determining surface and PBL fluxes from different types of mean flow and turbulence measurements
- Parameterizing turbulence and diffusion in the atmospheric surface layer
- Determining the surface energy budget

PROBLEMS AND EXERCISES

1. What maximum error or uncertainty might be expected in the determination of the Obukhov length, if the fluxes of momentum and heat have uncertainties of $\pm 20\%$ associated with them?

2. Using the definitions of the various scales and parameters and the basic Monin–Obukhov similarity relations [Eq. (11.2)], verify or derive Eqs. (11.4) and (11.5).

3. Using the forms of similarity functions obtained in the 1968 Kansas Field Program, calculate and plot the Richardson number as a function of z/L in the range $-2 \leq z/L \leq 2$, and compare it with the graph based on the simpler relations [Eq. (11.10)]. What conclusions can you draw from this?

4. Substituting from the M–O similarity profile relations in the bulk transfer formulas [Eq. (11.15)]
   (a) Derive the expressions given by Eq. (11.16) for C_D and C_H as functions of z/L and z/z_0.
   (b) Write the same as functions of the bulk Richardson number Ri_B for stable conditions ($Ri_B > 0$), and plot C_D/C_{DN} as a function of Ri_B for $z/z_0 = 10^3$ and 10^6.
   What conclusions can you draw from this?

5. The following measurements of mean wind and potential temperature were taken around noon during the 1968 Kansas Field Program:

z (m)	2	4	8	16	32
U (m sec^{-1})	5.81	6.70	7.49	8.14	8.66
Θ (K)	307.20	306.65	306.28	305.88	305.62

Using the M–O similarity relations $\phi_h = \phi_m^2 = [1 - 15(z/L)]^{-1/2}$ for the unstable surface layer
(a) Calculate and plot Ri as a function of height.
(b) Determine the Obukhov length from the above plot.
(c) Calculate the surface shear stress τ_0 and the heat flux H_0, using the above data from the lowest two (2 and 4 m) heights; take $\rho = 1.2$ kg m^{-3} and $c_p = 10^3$ J kg^{-1} K^{-1}.

6. The following observations were taken on a summer evening during the Kansas Experiment (with a surface pressure of 1000 mbar):

z (m)	2	4	8	16
U (m sec^{-1})	2.84	3.39	4.01	4.85
T (°C)	33.09	33.31	33.57	33.82

(a) Use the graphical version of the profile method discussed in the text to determine u_* and θ_*.
(b) Use the gradient or aerodynamic method for the same purpose, using information from the lowest two (2 and 4 m) heights only.
(c) Compare the results of the above two methods with the eddy correlation measurements of $\overline{uw} = -0.044$ m^2 sec^{-2} and $\overline{w\theta} = -0.023$ m sec^{-1} K.

7. During the Minnesota 1973 Experiment, the following mean temperature measurements were made from a tethered balloon at 75-min intervals:

	T (°C)	
z (m)	1217–1332 (CDT)	1332–1447 (CDT)
2	21.75	22.46
61	19.54	20.25
305	17.16	17.86
610	14.43	15.16
914	11.66	12.34
1219	8.99	9.61

(a) Calculate the warming rate at various levels and the average value for the whole PBL.

(b) Calculate the surface heat flux in the middle of the observation period when the PBL height was 1400 m, assuming a constant (average) warming rate, independent of height. Compare this with the direct (eddy correlation) measurement of $\overline{w\theta} = 0.20$ m sec^{-1} K.

(c) Suggest a simpler method of estimating the surface heat flux in daytime, using temperature measurements at only one level and remote sensing of the PBL height from the surface.

Chapter 12 | Evaporation from Homogeneous Surfaces

12.1 THE PROCESS OF EVAPORATION

The evaporation in the atmosphere and consequent phenomena of condensation, cloud formation, and precipitation have fascinated scientists, philosophers, and common people since time immemorial. Early philosophers and scientists proposed many theories and explanations of the evaporation process; for a brief review of the history the reader may refer to Brutsaert (1982). As early as in the fourth century BC, Aristotle recognized that "wind is more influential in evaporation than the sun." However, it was not until the nineteenth century that the foundations of the modern quantitative theories of evaporation and other transport phenomena were laid by J. Dalton, A. Fick, and O. Reynolds.

Evaporation is the phenomenon by which a substance is converted from the liquid state into vapor state. In the atmospheric surface layer evaporation may take place from a free water surface or a moist soil surface, as well as from the leaves of living plants and trees (here, we are ignoring the distinction between evaporation and transpiration). The details of the physical process of conversion from the liquid to vapor state may differ in different cases, but the basic mechanisms of transport of vapor away from any interface are the same. Very close to the interface, vapor is transferred through molecular exchanges in the same manner as heat and momentum are transferred. Most of the transfer takes place within a few molecular free path lengths of the surface. The vigorous molecular activity in this region is indicated by the observation that an estimated 2 to 3 kg of water moves across a free water interface in each direction every second, for each square meter of the interface (Munn, 1966, Chap. 10). The net transport of water vapor is only a tiny fraction of the total transport in each direction. According to Fick's law, the net flux of a material (e.g., water vapor) in a given direction is proportional to its concentration (specific humidity) gradient in that direction, the coefficient of proportionality being defined as the moecular diffusivity. Thus, the evaporation rate at a horizontal surface is given by

$$E_0 = -\rho\alpha_w(\partial Q/\partial z)$$ (12.1)

in which α_w is the molecular diffusivity of water vapor. The above relationship is expected to be valid only in the laminar or viscous sublayer in which the molecular exchange remains the primary (perhaps the only) transport mechanism. The same mechanism is operating at the air–soil and air–leaf interfaces. However, natural surfaces, not being very smooth and uniform, may not have well-defined molecular sublayers which could be amenable to direct measurements. Therefore, the practical utility of Eq. (12.1) for estimating evaporation in the atmosphere remains questionable. Some attempts have been made by physical chemists and fluid dynamists to study evaporation and other transport processes on the molecular scale. Micrometeorologists are more interested, however, in the exchange processes over the much larger scales that are encountered in the atmospheric surface layer and the PBL.

In the lower part (surface layer) of the atmospheric boundary layer the air motion is almost always turbulent. Here, the water vapor is efficiently transported away from the interfacial molecular sublayer by turbulent eddies. Because transport and the intensity of turbulent mixing depend on the surface roughness, wind shear or friction velocity, and thermal stratification, the evaporation rate also depends on the above factors, as well as on the average specific humidity gradient. The frequently used gradient-transport relation for the same is

$$E = -\rho K_w(\partial Q/\partial z)$$ (12.2)

in which the eddy diffusivity K_w replaces the molecular diffusivity in Eq. (12.1).

While Fick's law or Eq. (12.1) has a firm theoretical (e.g., the kinetic theory of gases) and experimental basis, the analogous relation [Eq. (12.2)] for a turbulent flow is based on a phenomenological gradient-transport theory whose limitations have already been pointed out in Chapter 9. Better flux-gradient relations obtained on the basis of the Monin–Obukhov similarity theory and carefully conducted micrometeorological experiments will be discussed in a later section.

12.2 POTENTIAL EVAPORATION AND EVAPOTRANSPIRATION

Over a bare land surface with no standing water, the soil moisture is the only source of water for evaporation. Therefore, the rate of evaporation must depend on the moisture content of the topmost layer of the soil. When the soil surface is fully saturated and the soil moisture content is not a limiting factor in evaporation, this maximum rate of evaporation for the given surface weather conditions is called potential evaporation (E_p). When the soil becomes drier, the rate of evaporation for the given atmospheric conditions (particularly the surface temperature, near-surface wind speed, specific humidity, and stability) also depends on the moisture content of the topmost soil layer which, in turn, depends on the soil moisture flow through the soil. The relationship between the rate of evaporation (E_0) and the soil moisture flow rate (M) is given by the instantaneous water balance equation at the surface

$$E_0 = M_0 \qquad (12.3)$$

or the water balance equation for the subsurface layer,

$$E_0 = M_b - \Delta M \qquad (12.4)$$

in which ΔM is the rate of storage of soil moisture in the layer per unit area of the surface, and M_b is the moisture flow rate at the bottom of the layer. The above simplified water budget equations would be valid only during the periods of no precipitation, irrigation, or runoff at the surface.

The vertical moisture flow rate, or the flux of soil moisture, is related to the vertical gradient of the soil moisture content (S) through Darcy's law

$$M = -k_m(\partial S/\partial z) \qquad (12.5)$$

where k_m is the hydraulic conductivity through the soil medium. Note that Eq. (12.5) (Darcy's law) is analogous to Eq. (4.1) (Fourier's law) for heat conduction. Similarly, following the derivation of Fourier's heat conduction equation [Eq. (4.2) or (4.3)], one can obtain

$$(\partial/\partial t)(\rho S) = -\partial M/\partial z \qquad (12.6)$$

or, after substituting from Eq. (12.5),

$$\partial S/\partial t = (\partial/\partial z)[\alpha_m(\partial S/\partial z)] \qquad (12.7)$$

where α_m is the diffusivity of soil moisture.

For surfaces with live vegetation, a significant portion of water vapor transfer to the atmosphere is by transpiration from the vegetation, most of which takes place through stomata of leaves. Transpiration may be considered an important by-product of the process of photosynthesis and respiration, in which plants take CO_2 during daytime and give out the same at nighttime. Transpiration may also be a physiological necessity for plants to draw up dissolved nutrients from the soil through their roots. Moisture transfer through roots, stands, and leaves is considerably more efficient than that through the soil pores, because the former is forced by a strong osmotic pressure gradient. Still, evaporation from the bare soil surface in between the plants is by no means negligible. Because evaporation and transpiration occur simultaneously and it is not easy to distinguish between the vapor transferred by the two processes, the term evapotranspiration is sometimes used to describe the total water vapor transfer to the atmosphere. More often, though, meteorologists use evaporation as a substitute or synonym for evapotranspiration. The distinction between the two disappears, anyway, when one considers the vegetative surface as a composite of soil and leaf surfaces, which contribute to the total water vapor transfer (evaporation) to the atmosphere per unit horizontal area of the composite surface, per unit time. In this gross sense, details of the vegetation canopy (e.g., leaves and branches) and small-scale transfer processes around the individual surface elements are essentially glossed over and the canopy is considered as an idealized, homogeneous material layer which has a finite thickness and mass, and which can store or release heat and water vapor. In particular, the water budget for a vegetative surface is often too complicated to be used as a practical tool for estimating evaporation.

The concept of potential evaporation (E_p) can also be extended to any vegetative surface for which E_p represents the maximum evapotranspiration likely from the vegetative surface for a given set of surface weather conditions. The potential evapotranspiration is generally less than the free water surface evaporation under the same weather conditions, especially in humid regions. The former can exceed the latter, however, in certain arid regions and under certain weather conditions (see, e.g., Rosenberg *et al.*, 1983, Chapter 7).

Actual evaporation usually differs from the potential evaporation, because the surface may not be saturated and the plants may not be drawing water from the soil and transpiring at their maximum rate. Still, the concept of potential evaporation is quite useful in agricultural and hydrological applications.

12.3 MODIFIED MONIN–OBUKHOV SIMILARITY RELATIONS

The original Monin–Obukhov similarity hypothesis and subsequent relations based on the same, as discussed in the preceding chapter, are strictly valid when buoyancy effects of water vapor can be ignored. When substantial evaporation occurs and it affects the density stratificaiton in the surface layer, modified M–O similarity relations incorporating the buoyancy effects of water vapor would be more appropriate.

Some buoyancy effects of water vapor have already been discussed in Chapter 5, where the concepts of virtual temperature and virtual potential temperature were introduced. A similar concept is that of the virtual heat flux H_v, which may be interpreted as the flux of virtual temperature ($H_v = \rho c_p \overline{w\theta_v}$). The modified Monin–Obukhov similarity hypothesis states that, in a homogeneous and stationary atmospheric surface layer, the mean gradients and turbulence structure depend only on four independent variables: z, u_*, g/T_{v0}, and $H_{v0}/\rho c_p$. As mentioned in Chapter 5, the difference between the virtual temperature and actual temperature is no more than 7 K and frequently less than 2 K, so that the buoyancy parameter g/T_{v0} does not differ much from g/T_0. However, the virtual heat flux may differ significantly from the actual heat flux, and the two are approximately related as

$$H_v \simeq H + 0.61 c_p \Theta E \qquad (12.8)$$

This follows from the approximate relationship between the fluctuations of virtual temperature, actual temperature, and specific humidity

$$\theta_v \simeq \theta + 0.61 \Theta q \qquad (12.9)$$

which, in turn, can be derived from the relationship between $\tilde{\Theta}_v$ and $\tilde{\Theta}$, noting that $\tilde{\Theta}_v = \Theta_v + \theta_v$ and $\tilde{\Theta} = \Theta + \theta$. These derivations are left as an exercise for the reader.

The relation given by Eq. (12.8) can also be expressed in terms of the latent heat flux H_L, or the Bowen ratio B, as

$$H_v = H + a_\theta H_L = H(1 + a_\theta B^{-1}) \qquad (12.10)$$

in which $a_\theta = 0.61 c_p \Theta / L_e$ is a dimensionless coefficient. Since a_θ is only weakly dependent on the actual temperature in °C, a constant value of $a_\theta = 0.07$, corresponding to $\Theta = 280$ K, is frequently used in the above relations. From Eq. (12.10) it is obvious that the suggested modification to the M–O hypothesis is necessary only when $|B| < 1$, i.e., when the latent heat flux exceeds the sensible heat flux. This is usually the case over the oceans, but not so common over the land areas.

Note that in the modified similarity hypothesis both the sensible heat flux and water vapor flux are considered together in an appropriate combination (the virtual heat flux) and not separately. Therefore, a modified Obukhov length is defined as

$$L = -u_*^3 / \left(k \, \frac{g}{T_{v0}} \, \frac{H_{v0}}{\rho c_p} \right) \qquad (12.11)$$

However, the scales of temperature, virtual temperature, and specific humidity have to be defined from their respective fluxes, i.e.,

$$\theta_* = -H_0/(\rho c_p u_*)$$

$$\theta_{v*} = -H_{v0}/(\rho c_p u_*) \tag{12.12}$$

$$q_* = -E_0/(\rho u_*)$$

The corresponding flux-profile relations for the transfer of water vapor in the atmospheric surface layer are

$$\frac{kz}{q_*} \frac{\partial Q}{\partial z} = \phi_w \left(\frac{z}{L}\right)$$

$$\frac{Q - Q_0}{q_*} = \frac{1}{k} \left[\ln \frac{z}{z_0} - \psi_w \left(\frac{z}{L}\right) \right] \tag{12.13}$$

$$E_0 = -\rho C_W U(Q - Q_0)$$

which are analogous to the heat transfer relations discussed in Chapter 11. Thus, the transfer of water vapor and other gaseous substances from the surface to the atmosphere or vice versa can be treated in the same way as the transfer of heat. The similarity between water vapor and heat transfer implies that

$$\phi_w(z/L) = \phi_h(z/L); \qquad \psi_w(z/L) = \psi_h(z/L)$$

$$K_w = K_h; \qquad C_W = C_H \tag{12.14}$$

Experimental verification of Eqs. (12.14) has been done in a few micro-meteorological studies. For example, simultaneous determinations of $\phi_h(z/L)$ and $\phi_w(z/L)$ from micrometeorological observations at Kerang, Australia, are compared in Fig. 12.1. Note that there is no systematic and significant difference between these two functions, although there is more scatter in the estimates of $\phi_w(z/L)$, probably due to larger uncertainties in

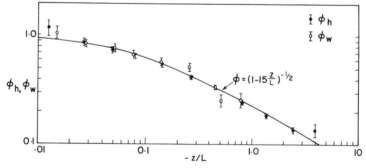

Fig. 12.1 Comparison of the observed dimensionless potential temperature and specific humidity similarity functions for unstable conditions. [After Dyer (1967).]

the eddy correlation measurements of $\overline{wq}$ as compared to $\overline{w\theta}$. Both of the empirical functions are well represented by the simpler Businger–Dyer relations [Eq. (11.9)]. A number of air–sea interaction studies in the marine atmospheric surface layer have also confirmed the approximate equality of eddy diffusivities, as well as bulk transfer coefficients of heat and water vapor, especially in the absence of such complicating factors as breaking waves and water spray.

12.4   MICROMETEOROLOGICAL METHODS OF DETERMINING EVAPORATION

Depending on the application for which the rate of evaporation is to be estimated and the associated temporal and spatial scales, a large number of hydrological, climatological, and micrometeorological methods have been proposed in the literature for determining evaporation (see, e.g., Brutsaert, 1982, Chapters 8–11; Rosenberg et al., 1983, Chapter 7). Here, we will restrict ourselves mainly to the micrometeorological methods which can be used to determine or measure the rate of evaporation at small time (order of an hour) and space (order of 100 m) scales.

12.4.1   DIRECT MEASUREMENTS BY LYSIMETERS AND EVAPORATION PANS

Instruments used for direct measurement of evaporation or potential evaporation from land surfaces are called lysimeters. A lysimeter is a cylindrical container (1–2 m deep and 1–6 m in diameter) in which an undisturbed block of soil and vegetation is isolated and its water budget is carefully monitored and controlled. Changes in the mass of the lysimeter are monitored either by a sensitive balance (mechanical or electronic) installed underneath, or a manometer measuring differences of hydrostatic pressures. There are two types of lysimeters, namely, weighing and floating lysimeters. In the latter, representative samples are floated on liquids such as water, oil, or solutions of zinc chloride, and their weight changes are determined from the fluid displacements resulting from the changes in buoyancy of the floating sample. Precision lysimeters with accuracies of 0.01–0.02 mm equivalent of water evaporated are found to be most suitable for measuring evaporation on short time scales (less than an hour). The lysimeter dimensions depend on the type of vegetation on the surface and the depth of the root zone. An extreme example is that of a giant floating lysimeter containing a mature Douglas fir tree in a forest at Cedar River, Washington, which is used for measuring evapotranspiration from the tree.

Evaporation pans are used for measuring free water evaporation from small lakes and ponds. Many types of pans have been used over the years and some have been standardized. The diameters of cylindrical evaporation pans used in practice range from 1 to 5 m and their depths range from 0.25 to 1.0 m. Due to limitations of size and spurious edge effects, the evaporation rate from a pan is usually 5–30% larger than that from a small lake or a pond. Large, sunken pans are most representative of free water evaporation in the absence of breaking waves and spray.

Evaporation pans have also been used for estimating potential evapotranspiration E_p from vegetative surfaces. Since, evaporation from a pan depends on its size, location, and exposure to sun and winds, calibration of a pan against measured potential evapotranspiration may be necessary for this method to be reliable. The ratio of potential evapotranspiration to pan evaporation on time scales of 1–30 days has been found to vary between 0.5 and 1.5 (this range may be expected to be even larger for small time scales of interest in micrometeorology). It depends on the pan size and exposure, geographical location, weather conditions, season, and the type of vegetation in its growth stage.

Other inexpensive instruments for estimating potential evaporation are atmometers, which are essentially porous ceramic or paper evaporating surfaces. The evaporating surface is continuously supplied with water; the measured rate at which water must be supplied to keep the porous material saturated is a measure of potential evaporation for given weather conditions. Again, calibration against a standard instrument or technique is necessary for an atmometer to be useful.

12.4.2 EDDY CORRELATION METHOD

A direct method of measuring the local water vapor flux over a homogeneous or nonhomogeneous surface is to measure simultaneously turbulent velocity and specific humidity fluctuations and determine their covariance over the desired sampling or averaging time. It is based on the relationship

$$E = \rho \overline{wq} \qquad (12.15)$$

where E is the vertical flux of water vapor. Over a homogeneous surface, if eddy correlation measurements are made in the constant flux surface layer, the rate of evaporation from the surface is also given by Eq. (12.15), as $E \simeq E_0$.

This method is simple in theory, but very difficult to use in practice, because the fast-response instruments required for measuring vertical velocity and specific humidity fluctuations need great care to install, maintain, calibrate, and operate. Reliable fast-response humidity instruments, such as the Lyman–Alpha humidiometer, microwave refractometer/hygrometer, and infrared hygrometer have been developed only recently. Such sophisticated instruments and the eddy correlation method have so far been used only during certain research expeditions and not for routine measurements of evaporation.

12.4.3   ENERGY BALANCE/BOWEN RATIO METHOD

The energy budgets of various types of surfaces, as well as of interfacial layers, have been discussed in Chapter 1. As mentioned in Chapter 11, the appropriate energy budget equation can be used to determine the sum of sensible and latent heat fluxes from measurements or estimates of the rest of the budget terms. Further partitioning of this into sensible and latent heat fluxes can be done if their ratio $B = H/H_L$ is measured or can be otherwise estimated. For example, from the surface energy budget equation [Eq. (2.1)] it follows that

$$H_0 = (R_N - H_G)/(1 + B^{-1})$$
$$E_0 = (1/L_e)(R_N - H_G)/(1 + B)$$

$$(12.16)$$

The Bowen ratio can be estimated from the gradient transport relations for H and E, with the assumption of the equality of the eddy exchange coefficients K_h and K_w. It is easy to show that

$$B = \frac{c_p}{L_e} \frac{\partial \Theta/\partial z}{\partial Q/\partial z} \simeq \frac{c_p}{L_e} \frac{\Delta \Theta}{\Delta Q} \tag{12.17}$$

where $\Delta\Theta = \Theta_2 - \Theta_1$ and $\Delta Q = Q_2 - Q_1$. Thus, in order to determine B one needs to measure the differences in temperature and specific humidities at two levels in the surface layer. This can be easily done through the use of dry- and wet-bulb thermometers or thermocouples, preferably in a difference circuit.

Another method of estimating the Bowen ratio is discussed later in the context of the Penman approach.

12.4.4   BULK TRANSFER APPROACH

This is similar to the bulk transfer method for determining the surface heat flux. The appropriate bulk transfer formula for the water vapor flux is

$$E_0 = \rho C_W U(Q_0 - Q) \tag{12.18}$$

in which Q_0 is the mean specific humidity very close to the surface (more appropriately, at $z = z_0$) and C_W is the bulk transfer coefficient for water vapor, which can be specified or parameterized in the same manner as C_H (Reynolds' analogy between water vapor and heat transfers implies $C_W = C_H$). This approach requires measurements of wind speed, temperature, and specific humidity at one level and knowledge of the surface roughness, temperature, and specific humidity. The main difficulty is that Θ_0 and Q_0 are not easy to measure or estimate, especially for vegetative surfaces.

12.4.5 PENMAN APPROACH

Using a combination of the energy balance and bulk transfer formulas, Penman (1948) derived a formula for evaporation from open water and saturated land surfaces, which considerably simplifies the observational procedure. Penman's formula, either in the original or slightly modified form, is widely used for estimating potential evaporation or evapotranspiration, especially in hydrological and agricultural applications. A modern derivation of the same is given in the following discussion.

Equation (12.18) can be written in the form

$$E_0 = \rho C_W U(Q_0 - Q_s) + E_a \qquad (12.19)$$

in which

$$E_a = \rho C_W U(Q_s - Q) \qquad (12.20)$$

can be interpreted as the drying power of air, because it is proportional to the difference between the saturation humidity and the actual specific humidity of air at the measurement height z.

If the measurement height is low (say, a few meters), the bulk transfer relation for the surface heat flux can be approximated as

$$H_0 = \rho c_p C_H U(T_0 - T) \qquad (12.21)$$

From Eqs. (12.19) and (12.21), assuming $C_W = C_H$, we obtain

$$\frac{L_e E_0}{H_0} = \frac{L_e}{c_p}\left(\frac{Q_0 - Q_s}{T_0 - T}\right) + \frac{L_e E_a}{H_0}$$

or, after converting specific humidities into water vapor pressures through Eq. (5.9),

$$B^{-1} = \frac{0.622 L_e}{P c_p}\left(\frac{e_0 - e_s}{T_0 - T}\right) + B^{-1}\frac{E_a}{E_0}$$

From this one obtains Penman's expression for the Bowen ratio

$$B = \frac{\gamma}{\Delta}\left(\frac{E_0 - E_a}{E_0}\right) \qquad (12.22)$$

where

$$\gamma = \frac{c_p P}{0.622 L_e} \qquad (12.23)$$

is the psychrometer constant whose value is about 0.66 mbar K^{-1} at $T =$ 20°C and $P = 1000$ mbar and

$$\Delta = (e_0 - e_s)/(T_0 - T) \simeq de_s/dT \qquad (12.24)$$

is the slope of the saturated vapor pressure versus temperature curve at T. The crucial approximation in Eq. (12.24), originally suggested by Penman (1948), could be justified only when the surface is wet, so that the water vapor pressure e_0 is close to its saturation value at the surface temperature T_0, and the measurement height above the surface is low.

After substituting from Eq. (12.22) into the second part of Eq. (12.16), we obtain the well-known Penman (1948) formula

$$E_0 = \left(\frac{\Delta}{\Delta + \gamma}\right)\left(\frac{R_N - H_G}{L_e}\right) + \frac{\gamma}{\Delta + \gamma} E_a \qquad (12.25)$$

It is clear from the above derivation that Penman's formula has a sound theoretical and physical basis, although some investigators have considered it empirical. In order to determine the evaporation rate from Eq. (12.25), one needs measurements or estimates of net radiation and soil heat flux at the ground level, together with the measurements of wind speed, temperature, and humidity at a low level (usually, 2 m) above the surface. The observational procedure is simplified by the ingenuous way in which the surface temperature has been eliminated from the final result.

Several further simplifications of Penman's approach to determining potential evaporation have been suggested in the literature. For example, over a water surface or a wet land surface the air might be close to saturation in water vapor, so that E_a would be expected to be negligibly small in comparison to E_0. Then, the simplified relation

$$E_0 = \frac{\Delta}{\Delta + \gamma} \left(\frac{R_N - H_G}{L_e}\right) \qquad (12.26)$$

should be adequate for determining the potential evaporation or evapotranspiration. The above formula was first suggested by Slatyer and McIlroy (1961), who called the left-hand term in Eq. (12.26) "equilibrium evapotranspiration." It has been observed that the simpler formula [Eq. (12.26)] is valid over a wide range of meteorological conditions and even for moderately dry (not fully saturated) surfaces with vegetation. Note that the only meteorological observation required is that of mean temperature at a low height of about 2 m, in addition to the measurement or estimate of $R_N - H_G$.

When Eq. (12.26) is valid, the Bowen ratio is simply estimated as

$$B \simeq \gamma/\Delta \tag{12.27}$$

which also follows from Eq. (12.22) when $E_a \ll E_0$. Equation (12.27) can be used in conjunction with Eq. (12.26) to obtain the sensible heat flux as

$$H_0 \simeq [\gamma/(\Delta + \gamma)](R_N - H_G) \tag{12.28}$$

Equations (12.26) and (12.28) are most suitable for determining the surface fluxes of water vapor and heat from routine micrometeorological measurements. Their applicability is, however, limited to wet, bare surfaces and vegetative surfaces over which evapotranspiration is near its potential rate and both H_0 and E_0 are positive (upward) fluxes. For water surfaces and those with tall vegetation, the use of the surface energy balance equation [Eq. (2.1)] may not be appropriate. For these, the energy budget for a layer can be used instead, and Eqs. (12.16), (12.25), (12.26), and (12.28) can be modified accordingly (replacing $R_N - H_G$ with $R_N - H_G - \Delta H_S$).

12.4.6 GRADIENT OR AERODYNAMIC METHOD

This is similar to the gradient-transport method of determining surface heat flux in Chapter 11. It requires measurements of specific humidity, in addition to those of wind speed and temperature, at two levels in the surface layer. Under neutral stability conditions, both the wind speed and specific humidity follow similar logarithmic profile laws. It is easy to show that the evaporation rate is given by the Thornthwaite–Holtzman forumla

$$E_0 = \rho k^2 \Delta U \Delta Q/[\ln(z_2/z_1)]^2 \tag{12.29}$$

For stratified conditions, following the same procedure as in the determination of momentum and heat fluxes in Chapter 11, the water vapor flux is given by

$$E_0 = -\rho k^2 \Delta U \Delta Q/\phi_m(\zeta_m)\phi_w(\zeta_m)[\ln(z_2/z_1)]^2 \tag{12.30}$$

which may be considered as a generalization of Eq. (12.29). Alternatively, one can use the gradient-transport relation [Eq. (12.2)] with $K_w = K_H$. The eddy diffusivity may be specified on the basis of Monin–Obukhov similarity relations, as described in Chapter 11.

12.4.7 PROFILE METHOD

Again, this is similar to the profile method of determining momentum and heat fluxes. The additional profile relationship to be fitted through the specific humidity data is

$$Q = (q_*/k)[\ln z - \psi_w(z/L)] + [Q_0 - (q_*/k) \ln z_0] \qquad (12.31)$$

which suggests a plot of Q versus $\ln z - \psi_w(z/L)$ for determining q_* and, if desired, Q_0. The friction velocity u_* is determined from a similar plot of wind profile and, then, the rate of evaporation $E_0 = -ku_*q_*$.

12.5 APPLICATIONS

The material presented in this chapter may have the following practical applications:

- Estimating evaporation or evapotranspiration over various types of surfaces
- Parameterizing the transfer of water vapor to the atmosphere in large-scale atmospheric models
- Systematic ordering of specific humidity observations in the atmospheric surface layer
- Determining the water budget of the earth's surface
- Incorporating the buoyancy effects of water vapor in flux-profile relations in the lower atmosphere

PROBLEMS AND EXERCISES

1. (a) Considering the flow of moisture only in the vertical direction through a uniform and homogeneous subsurface stratum, derive Eq. (12.7) for the soil moisture content.
   (b) Describe a plausible method of estimating the rate of evaporation from a bare soil surface, using measurements of the soil moisture content in the topmost layer.
2. (a) Derive an expression for the virtual heat flux in terms of the direct and latent heat fluxes.
   (b) If evaporation is taking place from a sea surface of temperature 27°C, which is slightly cooler than the air at a 10-m height, what value of the Bowen ratio would correspond to the neutral stability? What would be the corresponding heat flux if the rate of evaporation is 2 mm day^{-1}?
3. (a) If the normalized potential temperature and specific humidity gradients in the surface layer are equal [i.e., $(kz/\theta_*)(\partial\Theta/\partial z) = (kz/q_*)(\partial Q/\partial z)$], show that the eddy diffusivities of heat and water vapor must also be equal (i.e., $K_h = K_w$).

(b) How would you expect the ratio K_w/K_m to vary with stability? Express this ratio as a function of Ri.

(c) Calculate and compare the values of eddy diffusivity K_h or K_w during the typical daytime unstable ($u_* = 0.5$ m sec^{-1}, $H_{v0} = 500$ W m^{-2}) and nighttime stable ($u_* = 0.1$ m sec^{-1}, $H_{v0} = -50$ W m^{-2}) conditions, at a height of 10 m over a bare soil surface.

4. Starting from the instantaneous vertical flux of water vapor, derive the eddy correlation formula for the rate of evaporation $E_0 = \rho \overline{wq}$, and state all the assumptions you have to make.

5. Compare and contrast the energy balance/Bowen ratio method with the bulk transfer method for estimating heat and water vapor fluxes over (a) a short grass surface and (b) an ocean surface. Which method would you recommend in each case?

6. The following micrometeorological measurements of mean wind speed, temperature, specific humidity, and the surface energy budget were made on October 7, 1967, 1200 PST, on a large grass field near Davis, California:

z (m)	U (m sec^{-1})	T (°C)	Q (g kg^{-1})
0.25	1.11	23.52	8.22
0.50	1.34	23.35	7.73
1.00	1.51	23.30	6.88
2.00	1.63	23.14	6.65
6.00	1.80	22.95	6.05

Net radiation flux = 450 W m^{-2}
Ground heat flux = 43 W m^{-2}
Average grass height = 0.07 m

(a) Determine the average value of the Bowen ratio.

(b) Calculate the sensible heat flux and the evaporation rate, using the surface energy balance/Bowen ratio method.

(c) Calculate the potential evapotranspiration from the grass surface using the simplified Penman formula [Eq. (12.26)] with the observed air temperature at 2 m.

(d) Do the same, but using the original Penman formula [Eq. (12.25)] and taking the neutral value of $C_W = C_D = k^2/(\ln z/z_0)^2$.

7. Given the profile data of Problem 6, calculate the sensible heat flux and the rate of evaporation using the following methods:

(a) The gradient or aerodynamic method with observations at 0.5- and 2-m heights

(b) The profile method

Chapter 13 | Marine Atmospheric Boundary Layer

13.1 SEA-SURFACE CHARACTERISTICS

Over two-thirds of the earth's surface is composed of water in the form of oceans, seas, and lakes. Important exchanges of energy, mass, and momentum occur across these water surfaces, which influence atmospheric and oceanic circulations over a whole spectrum of time and spatial scales. Here, we will confine ourselves to the small-scale air–sea interaction processes that greatly influence the marine atmospheric boundary layer, as well as the upper oceanic mixed layer. Important differences between the PBLs over land and ocean surfaces arise because of some special thermodynamic and dynamic characteristics of the latter.

13.1.1 SEA–SURFACE TEMPERATURE AND ENERGY BUDGET

The most important characteristic of the sea surface, so far as energy exchanges across the interface are concerned, is its temperature. As distinguished from most land surfaces, open sea and ocean surfaces are characterized by a remarkable temporal and spatial homogeneity of temperature, particularly in the range of scales of interest in micrometeorology. This is primarily due to the large heat capacity and efficient mixing processes in the upper oceanic mixed layer.

The sea-surface temperature (T_s) depends on a number of oceanic, atmospheric, and other factors. Among the oceanic factors are the mixed-layer depth, the intensity of turbulent mixing, the presence of upwelling or downwelling, and the advective heat transport by oceanic currents. The meteorological factors affecting the sea-surface temperature are the net radiation to or from the sea surface, evaporation and precipitation processes, and, to a lesser extent, the sensible heat exchange with the atmosphere. The connection of all of these factors to the surface temperature becomes clear when one considers the thermodynamic energy balance at the sea surface. A simplified version of this was briefly discussed in Chapter 1.

Radiation balance near the sea surface is greatly complicated by the fact that the solar radiation received at the surface can penetrate to considerable depths of water and the shortwave reflectivity (albedo) of the surface is quite sensitive to solar altitude (see Munn, 1966, Chap. 18). The longwave radiation from the atmosphere is essentially absorbed by a very thin surface film of water which also gives out radiation to the atmosphere, depending on the surface temperature. The net longwave radiation depends on the presence of clouds and fog; it may vary with height above the surface due to absorption and emission by water vapor. It is the most important source of energy to the atmosphere.

A major component of the energy balance at the sea surface is the latent heat of evaporation (H_L), which is usually an order of magnitude larger than the sensible heat flux. Evaporation from ocean surfaces is not only the major source of water in the atmosphere, but it is also an important source of energy (made available at the time of condensation) that drives the atmosphere. It occurs most of the time, day and night.

The direct or sensible heat exchange (H) between the atmosphere and ocean is usually much smaller than the latent heat and radiative exchanges. It depends to a large extent on the air–sea temperature difference ($T_{10} - T_s$), which typically remains within ± 1 K over most open seas and oceans. During brief episodes of cold air outbreaks over warmer seas, oceans, and ocean currents, however, air–sea temperature differences can easily exceed 5 K. The sensible heat flux from the sea surface to the atmosphere becomes a significant part of the energy balance during such episodes. In extreme cases, air–sea temperature differences of 25–30 K have been observed (e.g., at the beginning of a severe cold air outbreak over the warm Gulf Stream, near Cape Hatteras, during the 1986 Genesis of Atlantic Lows Experiment), with the most spectacular signs (e.g., sea smoke, steam devils, and water spouts) of intense convection. Under such conditions, both the direct and latent heat fluxes become very large and nearly of the same order of magnitude [B = $H/H_L \sim 0(1)$].

The most uncertain part of the energy balance at the sea surface is the heat exchange H_G through the motions in water. Some of it is by turbulent transfer and mixing, which depends on the temperature difference across the oceanic mixed layer. Near-surface currents can also transport heat in the horizontal direction, and upwelling and downwelling motions transport in the vertical direction. In order to get around the difficulty of measuring or estimating H_G and also to account for the penetration of solar radiation through the depth of the mixed layer, it is found more convenient to consider the energy balance of the whole mixed layer

$$R_N = H + H_L + \Delta H_S \qquad (13.1)$$

in which the rate of heat storage ΔH_S is given by Eq. (2.3) and can be evaluated from temperature measurements at several depths in the mixed layer.

Because each term in Eq. (13.1) depends on the sea-surface temperature T_s, after appropriate parameterization of the various terms Eq. (13.1) can be used as a predictive equation for T_s. Such a procedure is rarely used in practice, however, because sea-surface temperatures are routinely and regularly monitored by environmental and earth resources satellites. The accuracy of such measurements, at present, is no better than $\pm 0.5°C$ and needs to be improved for more reliable and routine estimates of air–sea energy exchanges over large oceanic areas. The accuracy of radiometric measurements of the sea-surface temperature from low-flying research aircraft and ships can be better than $\pm 0.2°C$. When near-surface water temperature is measured with an *in situ* thermistor or thermometer attached to a float to keep it submerged, or after drawing the water sample, using a bucket, aboard a ship or boat, one should keep in mind the distinction between the sea-surface temperature and the subsurface or bucket temperature. Observations have shown that the difference between the surface temperature, as measured by a downward-looking radiometer, and the bucket temperature is typically $-0.5°C$ and can be as large as $-3°C$ during extreme free-convection conditions. The surface is generally colder than the water in the bulk of the mixed layer because of the combined influence of evaporative cooling at the surface and radiative warming of the whole mixed layer. The sharpest temperature gradient (up to $1°C \ mm^{-1}$) occurs in the topmost molecular (viscous) sublayer in which heat is transferred primarily through conduction.

The diurnal variation of T_s mainly depends on the diurnal changes of net radiation R_N and, to a lesser extent, of the other components of the surface energy budget. Despite the large diurnal changes in R_N, however, the diurnal range of the sea-surface temperature remains small (typically, less than $0.5°C$), because of the large heat capacity of the oceanic mixed layer. In shallow coastal waters, however, diurnal variations of $2–3°C$ in T_s can occur. Note that due to convective heat transfer through the bulk of the mixed layer, its depth is one to two orders of magnitude larger than the damping depth of diurnal influence in a soil medium.

Observations of the diurnal variation of T_s from research vessels indicate that the minimum temperature occurs in the morning hours before sunrise and the maximum value occurs in the afternoon. The actual times of minimum and maximum and the range of diurnal variation of the sea-surface temperature depend on the latitude, season, and the prevalent weather conditions at the location of interest. The diurnal range, $\Delta T_s = T_{smax} - T_{smin}$, is found to decrease with increasing latitude, wind speed, and cloudiness. The typical values of ΔT_s are 0.5°C in the tropics, 0.25°C in midlatitudes, and less than 0.1°C in high-latitude regions. Similar variations in T_s during the course of a day also occur intermittently when major storms pass through the area. For example, a marked rise of sea-surface temperature usually occurs during the approach of a tropical storm, hurricane, or typhoon; it is followed by a similar fall in the surface temperature after the passage of the storm center.

13.1.2 SURFACE WAVES

The most interesting and fascinating aspect of the sea surface is its endless, unceasing, up and down movement in the form of waves—a succession of humps and hollows in ever-changing patterns. Sea-surface waves are generated by atmospheric winds, astronomical causes (e.g., high and low tides), seismic disturbances, and, on a small scale, by passing ships and boats. The most common and persistent of these are wind-generated waves. The short waves generated and maintained by local winds are called seas, while long waves generated elsewhere and propagated into the region of interest are called swells. The former are attenuated rapidly as the local winds subside, while the latter continue even during long periods of near-calm conditions, because they are generated by distant forces and do not dissipate easily. Surface waves are also classified according to water depth (e.g., shallow-water and deep-water waves) and also on the basis of the governing force (e.g., gravity waves and capillary waves).

A regular (periodic) wave propagating at a constant velocity C_w can be characterized by its height H (vertical distance between a trough and the following crest), wave period T (the time elapsed between the passage of two successive crests or troughs past a fixed observer), and wavelength L (the horizontal distance between the adjacent crests or troughs at any instant). Alternatively, one can use the wave frequency $n = T^{-1}$ (the number of waves per unit time) or the wave number $\kappa = L^{-1}$ (the number of waves per unit length). For our simple wave form, these are interrelated as

$$\kappa = n/C_w$$

$$C_w = L/T$$

The ratio H/L is called the wave steepness, which is only about half the angle from horizontal made by a line from trough to crest.

Wind-generated ocean-surface waves are quite irregular, nearly chaotic, and three dimensional. Successive waves differ markedly in height, period, speed, and direction of propagation. Using the simplest statistics, one can determine from a record of surface elevation (η) the mean wave height $\bar{H}$, the root-mean-square of surface elevation σ_η, or the significant wave height $\bar{H}_{1/3}$ (defined as the mean of the one-third highest waves in a record), as well as the corresponding wave period. These average wave characteristics are essentially determined by large (dominant) waves and are not very sensitive to the hierarchy of smaller but steeper waves and wavelets that ride on the large waves.

More sophisticated statistical analyses of an irregular wave field assume the field to be composed of a large number of elementary sinusoidal waves of different frequencies and wave numbers, and sort out the relative contribution of each to the total variance of surface elevation. A plot of the component variance versus frequency (or wave number) is called spectrum. The peak in spectrum gives a clear-cut indication of which frequency (or wave number) waves are most dominant in a given wave field. For illustrative purposes some observed wave spectra are shown in Fig. 13.1, indicating a strong dependence of wave spectrum on fetch (distance from the shore) along wind.

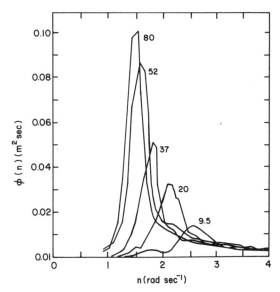

Fig. 13.1 A sequence of wave frequency spectra measured at increasing fetch (from 9.5 to 80 km) during the JONSWAP Experiment. [From Phillips (1977); after Hasselmann *et al.* (1973).]

When a light breeze starts blowing on a calm sea surface, the immediate first response of the water surface is the appearance of small, ruffled patches called cat's paws. This is in response to the small-scale turbulence, particularly pressure fluctuations, in the immediate vicinity of the surface. Then, small wavelets only a few millimeters in height and a few centimeters in wavelength appear. These tiny wavelets are quite steep and wrinkle the sea surface much more than larger waves do. Because this wrinkling is mainly resisted by surface tension associated with the capillary effect on liquids, such waves are called capillary waves (gravity is negligible in comparison to surface tension as a restoring force). If the breeze that initiated the capillary waves dies out, the waves also disappear quickly and the sea surface becomes smooth and glasslike again. However, if the breeze continues to blow, the capillary waves grow in height, as well as in wavelength, until they reach a wavelength of 17.2 mm, corresponding to a theoretical minimum wave speed of 0.231 m sec^{-1}.

For surface waves of larger wavelength, the gravity force becomes significant and even the dominant restoring force, so that such waves are called surface gravity waves. Unlike capillary waves, gravity waves do not dissipate easily, even long after the wind dies out, but continue to propagate for long distances from the region of their genesis. Thus, seas at one place eventually become swells at other places.

The sea state at any location and time depends on the strength and duration of winds, the fetch or distance over which winds blow in a coherent or persistent manner, and the strength and direction of surface currents. In coastal and shallow continental shelf waters, the wave field also depends on such factors as water depth, bottom slope, and the orientation of the coastline in relation to wind direction. The actual mechanisms and processes of generation, growth, and breaking of wind waves are too complex to be covered here. Essentially, they involve transfers of momentum and kinetic energy from the atmospheric surface layer to the oceanic surface layer.

During persistent stormy weather conditions over open oceans, waves can grow to spectacular and frightening dimensions. Observers have recorded waves of over 20 and even 30 m in height! But, more frequently encountered seas have significant heights of less than 2 m and wave periods of less than 10 sec. The wavelength and phase speed of an elementary component gravity wave are related to the wave period as

$$L = (g/2\pi) \, T^2$$
$$C_w = (g/2\pi) \, T$$

$$(13.2)$$

However, these relations do not hold for the average properties of a complex interference pattern of waves arising from the superposition of a number of elementary sinusoidal waves. It may not be very sensible even to talk about wave speed of such a nonconservative pattern, in which no individual wave crest can be followed long enough, because it loses its identity. Swell is more nearly a two-dimensional, regular, and periodic wave whose wavelength, period, and speed are approximately related by Eq. (13.2). The period of swell may range from several seconds to minutes.

13.2   MOMENTUM TRANSFER TO THE SEA SURFACE

The transfer of momentum from the atmosphere to the ocean at their interface is of considerable importance to meteorologists, as well as to physical oceanographers. A part of this momentum goes into the generation of surface waves, while the remaining portion is responsible for setting up drift currents and turbulence in the upper layers of the ocean. The relative partitioning of momentum between these different modes of motion in water depends on such factors as the stage of wave growth, duration and fetch of winds, etc., and is not easy to determine. But the total momentum exchange can be determined from measurements in the atmospheric surface layer, well above the level of highest wave crests.

Air flow in the lowest layer of direct wave influence is rather complex and is not easily measured or theoretically predicted. Above a few significant wave heights, however, the mean flow becomes essentially horizontal and unidirectional. In this so-called fully developed surface layer, the "law of the wall" similarity (the logarithmic wind-profile law) is valid under neutral conditions and the Monin–Obukhov similarity relations are applicable under more general stratified conditions. First, we consider the neutral or adiabatic case in order to clearly bring out the influence of the dynamic water surface, as distinguished from that of the rigid land surface.

When one considers extending the "law of the wall" similarity to air flow over a moving water surface, the effects of both the surface drift current and surface waves should be taken into account. The surface current, which is typically 3–5% of the wind speed at 10 m, but can be much stronger in certain regions (e.g., the Gulf Stream, Kuroshio, and other ocean currents), can easily be taken into account by considering the air motion relative to that of surface water ($U = U_a - U_w$). The influence of moving surface waves on mean wind profile in the surface layer is much more difficult to discern. However, the assumptions of stationarity and horizontal homogeneity implied in the surface layer and PBL similarity theories are much better satisfied over large lake and ocean surfaces, as compared to most land surfaces.

13.2.1   SEA-SURFACE ROUGHNESS

Wind-profile measurements from fixed towers in shallow coastal waters and from stable floating buoys (ships are not suitable for this, as they present too much of an obstruction to the air flow) over open oceans do show the validity of the logarithmic law [Eq. (10.6)] under neutral stability conditions. However, the roughness parameter (z_0) determined from the same is found to be related to both the wind and wave fields in a rather complex manner and can vary over a very wide range (say, 10^{-6}–10^{-2} m).

There have been a number of semiempirical and theoretical attempts at relating z_0 to the friction velocity, as well as to the average height, the phase speed, and the stage of development of surface waves. The simplest and still the most widely used relationship is that proposed by Charnock (1955) on the basis of dimensional arguments

$$z_0 = a \; (u_*^2/g) \tag{13.3}$$

where a is an empirical constant. The basic assumptions implied in Charnock's hypothesis (z_0 is uniquely determined by u_* and g) are that the winds are blowing steadily and long enough for the wave field to be in complete equilibrium with the wind field, independent of fetch, and that the surface is aerodynamically rough.

Individual estimates of z_0 plotted as a function of u_*^2/g invariably show large scatter, because they usually pertain to different fetches and different stages of wave development and also because of large errors in the experimental determination of z_0. However, when the available data sets are block averaged, a definite trend of z_0 increasing in proportion to u_*^2/g does emerge. Figure 13.2 shows such block-averaged data compiled from some 33 experimental sets of wind profile and drag data (Wu, 1980). Note that Charnock's formula [Eq. (13.3)] is verified, particularly for large values of z_0 and u_*^2/g, and $a \simeq 0.018$.

The simple state of affairs assumed in Charnock's hypothesis rarely exists over lakes and oceans. More frequently, the wave field is not in equilibrium with local winds, but depends on the strength, fetch, and time history of the wind field. At low wind speeds, the surface can also become aerodynamically smooth, or be in a transitional regime between smooth and rough. A comprehensive treatment of sea-surface roughness at different development stages of wind-generated waves is given by

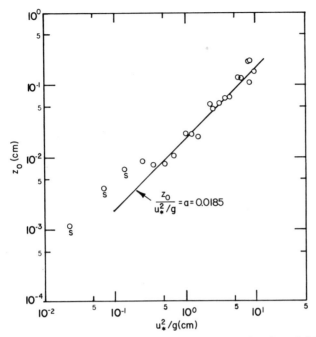

Fig. 13.2 Observed roughness parameter of sea surface as a function of u_*^2/g compared with Charnock's (1955) formula. [After Wu (1980).]

Kitaigorodski (1970). The fundamental parameter indicating the stage of wave development, in his theory, is the ratio C_0/u_*, where C_0 is the phase speed of the most dominant (corresponding to the peak in the wave-height spectrum) waves. It is shown that in the initial stage of wave development (very small C_0/u_*), the characteristic roughness height h_0 is proportional to σ_η, while in the equilibrium stage of wave development h_0 is proportional to u_*^2/g, as implied by Charnock's formula [Eq. (13.3)]. In a broad intermediate range of C_0/u_*, however, h_0 is shown to depend on both σ_η and C_0/u_*. Furthermore, on the basis of analogy with the aerodynamic roughness of a rigid surface, z_0 is assumed proportional to h_0 for a completely rough surface, proportional to ν/u_* for an aerodynamically smooth surface, and a function of both h_0 and ν/u_* in the transitionally smooth or rough regime. Kitaigorodski has suggested plausible expressions for the roughness parameter for the various combinations of roughness and wave regimes, some of them involving still unknown empirical constants or functions. In his matrix, Charnock's formula falls in one corner, corresponding to the aerodynamically rough, fully developed (equilibrium) wave regime.

If, for the sake of convenience and other practical reasons, Charnock's formula is used, irrespective of the wave development stage, one should expect $a = gz_0/u_*^2$ to be a function of at least C_0/u_*, if not any other parameter. Only in the asymptotic limit of $C_0/u_* \to \infty$, a may approach a constant value. This explains why empirically determined values of a by different investigators range from 0.01 to 0.08. Owing to the lack of simultaneous data on z_0 and sea state, the function $a(C_0/u_*)$ has not been well established, but observations indicate that $a = 0.015$–0.020 for the later (final) stages of wave development. For early stages of wave development, the following semiempirical formula suggested by Kitaigorodski (1970) can be recommended:

$$z_0 \simeq 0.12\bar{H} \exp(-kC_0/u_*), \qquad \text{for} \quad C_0/u_* \le 20 \qquad (13.4)$$

where $\bar{H} = (2\pi)^{1/2}\sigma_\eta$ is the mean wave height. The use of Eq. (13.4) would require, of course, detailed wave-height measurements. An experimental verification of Eq. (13.4) is shown in Fig. 13.3, which also demonstrates the large scatter in the individual estimates of z_0.

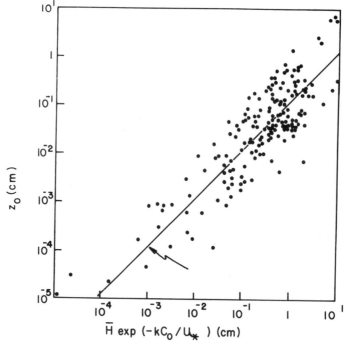

Fig. 13.3 Observed roughness parameter of sea surface as a function of the reduced wave height $H_0 \exp(-kC_0/u_*)$ compared with Eq. (13.4), indicated by the arrow. [After Kitaigorodski (1970).]

At low wind speeds (say $U_{10} < 7$ m sec^{-1}) and in early stages of wave development, neither Eq. (13.3) nor (13.4) may be adequate, because the surface is aerodynamically smooth or incompletely rough. The sea surface appears to be remarkably smooth, before waves start to break. In fact, the aerodynamic roughness of even a fully developed sea under a tropical storm is observed to be less than that of mown grass. This is largely due to the fact that waves are generated by and move along with the wind and do not offer as much resistance to the flow as would an immobile surface with similar shape. There is also strong evidence suggesting that sea-surface drag is primarily due to small ripples and wavelets with phase speeds less than the near-surface wind speed, and that large waves make only a minor contribution to the same.

Different surface roughness regimes mentioned in Section 10.1 are found to occur over oceans. In the absence of swell, the sea surface is found to be aerodynamically smooth during calm and weak winds ($U_{10} < 2.5$ m sec^{-1}), transitionally rough in moderate winds ($2.5 \leqslant U_{10} \leqslant 7.5$ m sec^{-1}), and completely rough in strong winds ($U_{10} > 7.5$ m sec^{-1}). The above criteria should be considered only approximate and limited to fully developed (equilibrium) wind-generated waves. The data points corresponding to small values of u_*^2/g in Fig. 13.2 probably belong to the transitional roughness regime.

13.2.2 SURFACE DRAG COEFFICIENT

The definition and concept of the surface drag coefficient was introduced in Section 10.3. It is found to be particularly useful in oceanographic and marine meteorological applications to express or parameterize the sea-surface drag in terms of wind speed at a standard height of 10 m and the drag coefficient C_D referred to that height. Note that under neutral stability conditions, C_D is uniquely related to the roughness parameter z_0 through Eq. (10.14). Still, most investigators prefer C_D over z_0, because of the much smaller range of variation of the former ($0.001 < C_D < 0.003$), as compared to the latter ($10^{-6} < z_0 < 10^{-2}$ m) over the oceans, and also because the surface stress is more directly related (proportional) to C_D, irrespective of the wind-profile shape. The drag coefficient can be determined from measurements of momentum flux and mean wind speed at only one level (usually 10 m), while a direct estimate of z_0 requires accurate wind-profile measurements in the surface layer. Both the drag and wind-profile measurements are quite sensitive to motions of the observation platform in response to surface waves, as well as to the deflection of flow around it.

There has been considerable interest in knowing the dependence of C_D

on wind speed, sea state, and atmospheric stability. To this end, many experimental determinations of C_D have been made by different investigators, using different measurement techniques. Early determinations were largely based on rather crude estimates of the tilt of water surface and the geostrophic departure of flow in the PBL, and are not considered very reliable. More recently, the sea-surface drag has been determined by using the more reliable eddy correlation, gradient, and profile methods. A comprehensive review of the existing data on C_D up to 1975 was made by Garratt (1977). After applying the appropriate corrections for atmospheric stability and actual observation height and using the Monin–Obukhov similarity relations, the neutral drag coefficient C_{DN}, referred to the standard height of 10 m, is obtained as a function of wind speed at 10 m (U_{10}).

Figure 13.4a shows the block-averaged values of C_{DN} ± 1 standard deviation (indicated by vertical bars) for intervals of U_{10} of 1 m sec^{-1}, based on the eddy correlation method (closed circles) and wind-profile method (open circles). The number of data used for averaging in each 1-m sec^{-1} interval are also indicated above the abscissa axis (the top line refers to closed circles and the bottom line to open circles). Note that there are too few data to give reliable estimates of C_{DN} at high wind speeds ($U_{10} > 15$ m sec^{-1}).

Some investigators have inferred the surface stress in hurricanes, using the geostrophic departure method. Others have measured C_{DN} at high wind speeds in laboratory wind flume experiments. Figure 13.4b shows that there is a remarkable similarity between the hurricane data and wind flume data (usually referred to a height of 10 cm, assuming a scaling factor of 100 : 1), despite the large expected errors in the former and dissimilarity of wave fields in the two cases.

Considering the overall block-averaged data for the whole wide range of wind speeds represented in Fig. 13.4a and b, there is no doubt that C_{DN} increases with wind speed. The trend appears to be quite consistent with Charnock's formula [Eq. (13.3)], which, in conjunction with Eqs. (10.13) and (10.14), yields

$$\ln C_{DN} + kC_{DN}^{-1/2} = \ln(gz/aU^2) \qquad (13.5)$$

After determining the best-fitting value of $a \simeq 0.0144$ with $k = 0.41$ (for the more commonly used value of $k = 0.4$, one would obtain $a \simeq 0.017$, which is closer to the value indicated in Fig. 13.2), from a regression of $\ln C_{DN} + kC_{DN}^{-1/2}$ on $\ln U_{10}$, Eq. (13.5) is as represented in Figs. 13.4a and b. In the completely rough regime ($U_{10} > 7$ m sec^{-1}), in particular, Charnock's relation is in good agreement with the experimental data. The variation of C_{DN} with U_{10} can also be approximated by a simple linear relation (Garratt, 1977)

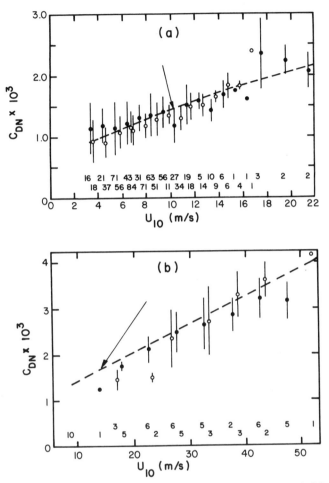

Fig. 13.4 Neutral drag coefficient as a function of wind speed at a 10-m height compared with Charnock's formula [Eq. (13.5), indicated by the arrows in (a) and (b)] with $a = 0.0144$. Block-averaged values are shown for (a) 1-m sec^{-1} intervals, based on eddy correlation and profile methods, and (b) 5-m sec^{-1} intervals, based on geostrophic departure method and wind flume simulation experiments. [After Garratt (1977).]

$$C_{DN} = (0.75 + 0.067U_{10}) \times 10^{-3} \qquad (13.6)$$

or a power relation

$$C_{DN} = 0.00051U_{10}^{0.46} \qquad (13.7)$$

in which U_{10} is to be expressed in meters per second.

A linear relationship, similar to Eq. (13.6), is also suggested by more recent measurements of drag in gale-force winds (see, e.g., Smith, 1980). Such linear or power relations, although convenient for practical applications, are not dimensionally homogeneous (empirical constants must have rather odd dimensions). Equation (13.5) suggests that the more appropriate dimensionless parameter governing C_{DN} is the Froude number $F \equiv U/(gz)^{1/2}$, which should be used instead of wind speed. For a fixed (standard) reference height of 10 m, however, F is directly proportional to the wind speed ($F \simeq 0.10U_{10}$, for U_{10} in meters per second) and the latter can be used for convenience.

There is some controversy over the variation of C_{DN} with U_{10} at low and moderate wind speeds (say, $U_{10} < 7.5$ m sec^{-1}). At very low wind speeds ($U_{10} < 2.5$ m sec^{-1}), observations from laboratory flume experiments and from over lakes and oceans indicate a slight tendency of C_{DN} to decrease with the increase in wind speed, as would be expected for an aerodynamic smooth surface. But sea surface is rarely completely smooth; ripples and wavelets generated by weak to moderate winds give it a moderate roughness, which is transitional between smooth and completely rough regimes. Some investigators have argued that the dependence of C_{DN} on U_{10} should be weak, if any, in this transitional regime ($2.5 < U_{10} < 7.5$ m sec^{-1}). The data of Fig. 13.4 do suggest a weaker dependence of C_{DN} on U_{10} in this range. Other oceanic data have even indicated a nearly constant value of $C_{DN} \simeq 1.2 \times 10^{-3}$ in the moderate range of wind speeds (see, e.g., Kraus, 1972).

Dependence of C_{DN} on other parameters, such as fetch, wind direction relative to the coastline, temporal changes in wind speed and direction, and stage of wave development, also need to be investigated. Most of the original studies of sea-surface drag do not provide quantitative information on these parameters and do not classify C_{DN} accordingly. Their effects remain obscured in the scatter of individual data sets around the mean values (indicated by standard deviations shown by vertical bars in Fig. 13.4a and b) and also in the scatter of mean values. Few investigators have sought to investigate the dependence of C_{DN} on wind direction, fetch, and sea-state parameters, such as C_0/u_* (see, e.g., Kitaigorodski, 1970; SethuRaman, 1978; Smith, 1980). There is only a slight tendency for the drag coefficient to decrease with increasing fetch. It has also been argued that the coefficient a in Charnock's formula should decrease with increasing C_0/u_* and may reach a constant asymptotic value only at the equilibrium stage of wave development. This would imply that C_{DN} should also decrease with increasing C_0/u_*. An observational evidence in support of such a trend is given by SethuRaman (1978), who also shows dramatic changes in C_{DN} in response to a rapid shift in wind direction.

13.3 PARAMETERIZATION OF AIR–SEA EXCHANGES

In addition to the transfer of momentum between the atmosphere and the ocean at their interface, substantial amounts of heat (both sensible and latent) and mass (e.g., water vapor, CO_2, O_2, and salt nuclei) are exchanged across the air–sea interface. Of primary importance to the dynamics and thermodynamics of the atmosphere are the exchanges of heat and water vapor, which are usually parameterized as

$$H_0 = \rho c_p C_H U(\Theta_0 - \Theta)$$

$$H_{L0} = L_e E_0 \tag{13.8}$$

$$E_0 = \rho C_W U(Q_0 - Q)$$

An experimental verification of Eq. (13.8) is given in Figs. 13.5 and 13.6, in which vertical fluxes of heat and moisture are shown to be well correlated (nearly proportional) with the products $U(\Theta_0 - \Theta)$ and $U(Q_0 - Q)$, respectively, at a reference height $z = 10$ m. Note that the bulk transfer coefficients C_H and C_W are given by the slopes of the regression lines passing through the origin in Figs. 13.5 and 13.6, respectively.

A central objective of many air–sea interaction experiments and expe-

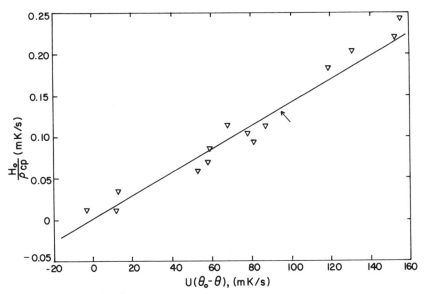

Fig. 13.5 Sensible heat flux at the sea surface as a function of $U(\Theta_0 - \Theta)$ in strong winds compared with Eq. (13.8) with $C_H = 1.41 \times 10^{-3}$, indicated by the arrow. [After Friehe and Schmitt (1976); data from Smith and Banke (1975).]

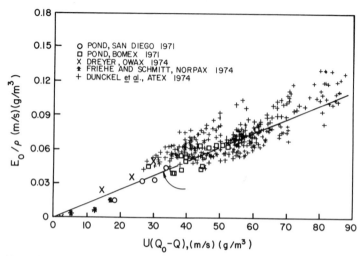

Fig. 13.6   Observed moisture flux at the sea surface as a function of $U(Q_0 - Q)$ compared with Eq. (13.8) with $C_W = 1.32 \times 10^{-3}$, indicated by the arrow. [After Friehe and Schmitt (1976).]

ditions has been the empirical determination of the drag, heat, and water vapor transfer coefficients (C_D, C_H, and C_W), covering a wide range of surface roughness, sea state, and stability conditions. Under moderate winds ($6 < U_{10} < 12$ m sec^{-1}) and near-neutral stability, there is an overwhelming experimental evidence indicating that $C_{DN} \simeq C_{HN} \simeq C_{WN} \simeq 1.2 \times 10^{-3}$. Under more typical, slightly unstable conditions prevailing over the oceans, however, transfer coefficients are somewhat larger, but still approximately equal to each other ($C_D \simeq C_H \simeq C_W \simeq 1.5 \times 10^{-3}$), at least within the estimated margin of error ($\pm 20\%$) of their determination (see, e.g., Pond, 1975). These results point out to a rather simple parameterization of air–sea fluxes in terms of wind speed, temperature, and specific humidity at the 10-m level and the sea-surface temperature.

Atypical or exceptional stability conditions may be encountered, of course, over certain oceanic areas and during certain weather conditions when C_D, C_H, and C_W may considerably differ from their above typical values. For example, in the areas and periods of intense cold-air advection over the warmer ocean, free convection may occur in which C_H and C_W are expected to become considerably larger than $C_D > 1.5 \times 10^{-3}$. On the other hand, under inversion conditions and light winds, all the transfer coefficients are expected to become much smaller than their neutral value.

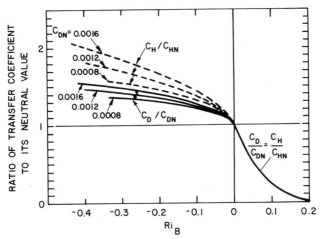

Fig. 13.7 Calculated ratios of the drag and heat transfer coefficients to their neutral values as functions of the bulk Richardson number and C_{DN}. [After Deardorff (1968).]

The stability effects on the drag and other transfer coefficients can be considered in the framework of the Monin–Obukhov similarity theory. For example, Eq. (11.16) can be expressed in the form

$$C_D/C_{DN} = [1 - k^{-1}C_{DN}^{1/2}\psi_m(z/L)]^{-2}$$
$$C_H/C_{DN} = [1 - k^{-1}C_{DN}^{1/2}\psi_m(z/L)]^{-1}[1 - k^{-1}C_{DN}^{1/2}\psi_h(z/L)]^{-1} \tag{13.9}$$

expressing the ratios C_D/C_{DN} and C_H/C_{DN} as functions C_{DN} and z/L. These can also be expressed in terms of the more convenient bulk Richardson number $Ri_B = gz(\Theta_v - \Theta_{v0})/T_{v0}U^2$, using the following relations between z/L and Ri_B:

$$z/L \simeq kC_{DN}^{-1/2}Ri_B, \qquad \text{for} \quad Ri_B < 0$$
$$z/L \simeq kC_{DN}^{-1/2}Ri_B(1 - 5Ri_B)^{-1}, \qquad \text{for} \quad Ri_B \geq 0 \tag{13.10}$$

The first of these relations is only an approximation to the more exact but implicit relationship between z/L and Ri_B. The above approximations are considered to be quite good for the expected range of Ri_B values ($-0.1 <$ $Ri_B < 0.1$) over the ocean. The ψ functions in Eq. (13.9) are given by Eq. (11.14), so that the ratios C_D/C_{DN} and C_H/C_{DN} can be evaluated as functions of Ri_B. Figure 13.7 shows that these are only weakly dependent on the neutral value of the drag coefficient (C_{DN}), but have rather strong dependence on Ri_B, particularly in stable conditions.

There are very few experimental studies in which stability effects on the drag and other transfer coefficients have been systematically investi-

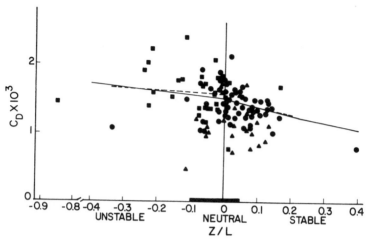

Fig. 13.8 Observed drag coefficient adjusted to a wind speed of 15 m sec^{-1} as a function of stability. The solid line shows expected dependence on stability from similarity flux-profile relations. ●, Long fetch; ■, limited fetch; ▲, alongshore. [After Smith (1980).]

gated (see, e.g., Smith, 1980). The limited range of stability parameter z/L or Ri_B encountered during any experiment and the inherent scatter in the data due to variations in fetch and duration of winds do not allow precise determinations of C_D, C_H, and C_W as functions of sea-surface roughness and stability (see, e.g., Fig. 13.8). All that can be said from such experimental evaluations is that they are not inconsistent with theoretical relations given above.

13.4 MEAN PROFILES IN THE MARINE ATMOSPHERIC BOUNDARY LAYER

Measurements of wind, temperature, and humidity profiles over the open ocean are made extremely difficult by undesirable motions of any floating platform and distortion of air flow around the same. Even more difficult to take and interpret are observations in the lowest layer of direct wave influence. Accurate profile measurements in the fully developed region of the atmospheric surface layer have been made from shallow-water towers and platforms, as well as from specially designed stable buoys and research vessels (see, e.g., Paulson *et al.*, 1972). These are found to be consistent and in good agreement with the Monin–Obukhov similarity relations [Eqs. (11.9)–(11.14)], provided the buoyancy effects of water vapor are included in the definitions of the M–O stability parame-

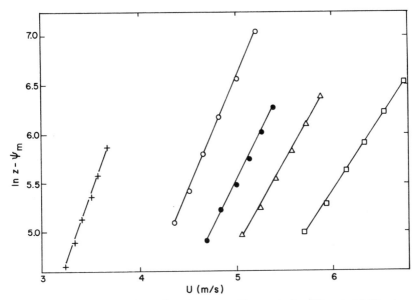

Fig. 13.9 Composite wind speeds as functions of ln $z - \psi_m$ for different stability classes during IIOE. Run: ○, I; □, II; △, III; ●, IV; +, V. [After Paulson (1967).]

ter, as described in Section 12.2. For example, Figs. 13.9 and 13.10 show average profiles for different stability classes, ranging from slightly stable (I) to moderately unstable (V), taken during the 1964 International Indian Ocean Expedition (IIOE). Note that according to the M–O similarity relations [Eq. (11.12)], plots of U versus ln $z - \psi_m$ and Θ versus ln $z - \psi_h$ are expected to be linear and are shown to be so. Also note the much smaller gradients of wind and temperature over the ocean, as compared to those over land surfaces, which limit the accuracy of flux determinations from mean profile measurements.

In the outer part of the marine atmospheric boundary layer (MABL), wind, temperature, and humidity profiles are usually taken from ship-based pilot balloons, radiosondes, minisondes, and more sophisticated profilers and sounding systems. In coastal regions, research aircraft may be used. Most sounding systems give only instantaneous or inadequately averaged winds and other meteorological variables, which generally deviate from and scatter around the expected mean profiles of these variables. Individual wind soundings, in particular, show considerable variability due to the presence of large eddies in the atmospheric boundary layer, and several of them taken under similar conditions may have to be averaged in order to get a smooth profile.

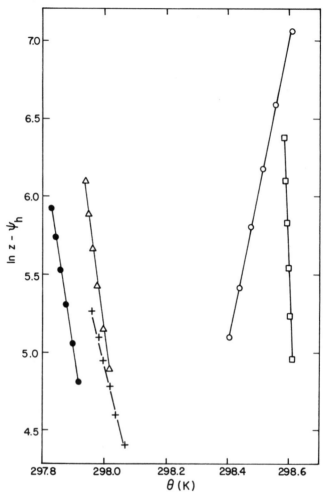

Fig. 13.10   Composite potential temperatures as functions of $\ln z - \psi_h$ for different stability classes during IIOE. Run: ○, I; □, II; △, III; ●, IV; +, V. [After Paulson (1967).]

Aircraft measurements can be used to obtain spatial averages along specified flight legs at different levels. For thermodynamic variables, differences between balloon soundings and spatially averaged profiles are usually small. Figure 13.11 shows a comparison of the two during the 1975 Air Mass Transformation Experiment (AMTEX) over the East China Sea. These profiles are typical of convective conditions prevailing during a cold-air outbreak.

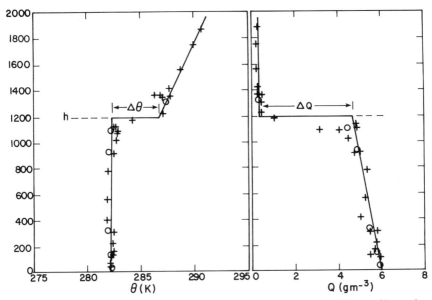

Fig. 13.11   Observed mean potential temperature and specific humidity profiles under convective conditions during AMTEX. ○, Leg average; +, sounding (February 15). [After Wyngaard *et al.* (1978).]

The potential temperature and specific humidity profiles in the MABL during more typical, slightly unstable, and fair-weather conditions are given in Fig. 13.12. These were measured by an aircraft during the 1974 GARP Atlantic Tropical Experiment (GATE). Note that in the subcloud layer, Θ, Θ_v, and, to some extent, Q profiles are nearly uniform. The

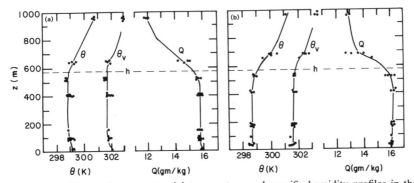

Fig. 13.12   Observed mean potential temperature and specific humidity profiles in the marine boundary layer under slightly unstable conditions during GATE. (a) Day 253 and (b) day 258. [After Nicholls and LeMone (1980).]

measured wind profiles during the same experiment showed a wide range of shapes and forms and are quite sensitive to stability, baroclinity, and presence of clouds in the boundary layer. In short, there is no typical wind profile for the MABL. Figure 6.6 is an example of wind component profiles in a trade-wind boundary layer.

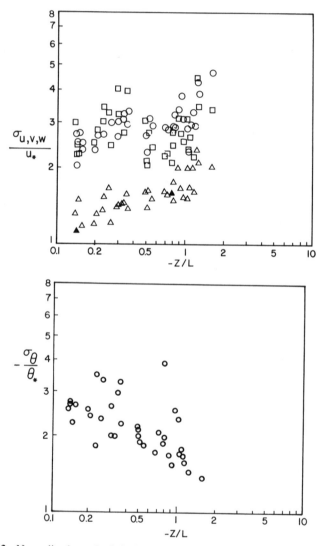

Fig. 13.13 Normalized standard deviations of velocity and temperature fluctuations in the marine surface layer as functions of stability during BOMEX. □, σ_u/u_*; ○, σ_v/u_*; △, σ_w/u_*. [After Leavitt and Paulson (1975).]

13.5   TURBULENCE OVER WATER

There is some observational evidence indicating that very close to the wavy sea surface, turbulence in air is modulated and affected by surface waves (see, e.g., Kitaigorodski, 1970). A few wave heights above the mean sea level, however, the direct influence of waves becomes insignificant and the fully developed surface layer turbulence is similar to that over homogeneous land surfaces. This means that the appropriate scales of velocity, temperature, and humidity fluctuations are u_*, θ_*, and q_*, while the ratio of the height scales z/L is the appropriate stability or similarity parameter.

The normalized standard deviations of velocity fluctuations in the surface layer, measured during the 1969 Barbados Oceanographic and Meteorological Experiment (BOMEX), are plotted against $-z/L$ in Fig. 13.13. Note that while σ_w/u_* apparently approaches the local free convection prediction $\sigma_w/u_* \sim (-z/L)^{1/3}$, for large values of $-z/L$, σ_u/u_* and σ_v/u_* do not show any such behavior. Other data over land and ocean surfaces also indicate that horizontal components of turbulence do not follow ei-

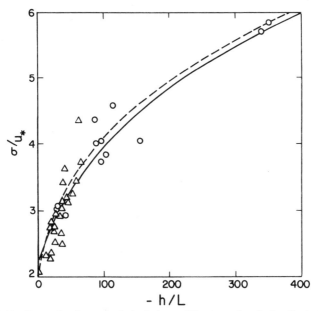

Fig. 13.14 Normalized standard deviations of horizontal velocity fluctuations in the surface layer as functions of h/L compared with mixed-layer similarity relations. ---, $\sigma/u_* = [4 + 0.6(-h/L)^{2/3}]^{1/2}$; ——, $\sigma/u_* = (12 - 0.5\,h/L)^{1/3}$; $\triangle$, airplane data; $\bigcirc$, Minnesota tower data. [After Panofsky *et al.* Copyright © (1977) by D. Reidel Company. Reprinted by permission.]

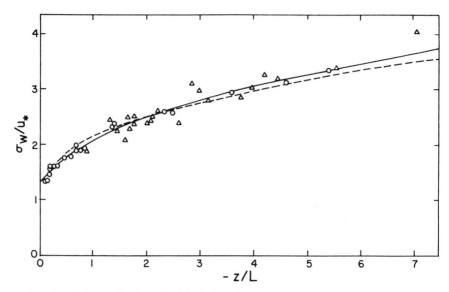

Fig. 13.15 Normalized standard deviation of vertical velocity fluctuations in the surface layer as a function of z/L compared with surface layer similarity relations. ---, $\sigma_w/u_* = [1.6 + 2.9(-z/L)^{2/3}]^{1/2}$; ——, $\sigma_w/u_* = 1.3[1 - 3(z/L)]^{1/3}$; $\triangle$, airplane data; $\bigcirc$, Minnesota tower data. [After Panofsky *et al.* Copyright © (1977) by D. Reidel Publishing Company. Reprinted by permission.]

ther Monin–Obukhov or local free convection similarity scaling, but they seem to follow the mixed-layer scaling, even in the surface layer, i.e., for $-h/L \gg 1$, $\sigma_u \sim W_*$ and $\sigma_v \sim W_*$. This implies that $\sigma_{u,v}/u_* \sim (-h/L)^{1/3}$. Figures 13.14 and 13.15 show different behaviors of horizontal and vertical components of turbulence and suggest different empirical relations for them in unstable and convective conditions.

13.6 APPLICATIONS

The material presented in this chapter may have applications in the following areas:

- Determining energy budgets of an ocean or sea surface
- Determining the air–sea exchanges of momentum, heat, and water vapor at the sea surface
- Predicting sea state, including storm surges
- Modeling atmospheric and oceanic boundary layers and the interaction between the two

- Modeling upper oceanic currents and circulations
- Forecasting storm tracks and weather at sea
- Long-range weather forecasting and climate simulations, using coupled upper-ocean–atmospheric models

PROBLEMS AND EXERCISES

1. The following observations were made during a period of intense cold air outbreak over the East China Sea during the 1974 Air Mass Transformation Experiment:

   Sensible heat flux from the sea surface = 220 W m^{-2}
   Latent heat flux from the sea surface = 490 W m^{-2}
   Net radiative loss from the sea surface = 30 W m^{-2}
   Sea-surface temperature   = 21.5°C
   Oceanic mixed-layer depth   = 100 m

   Calculate the rate of warming or cooling of the oceanic mixed layer in °C per day.

2. Calculate and compare the wavelengths and wave speeds of sea-surface swells of periods 10, 20, 30, 40, 50, and 60 sec.

3. Using Charnock's relation for the surface roughness and the logarithmic wind-profile law, derive Eq. (13.5) and plot C_{DN} as a function of the Froude number $F = U/(gz)^{1/2}$ in the wind speed range of 5 to 50 m sec^{-1} at $z = 10$ m.

4. (a) Starting from Eq. (11.16), derive Eq. (13.9).
   (b) Calculate and plot C_D/C_{DN} as a function of Ri_B in the range $-0.1 \leqslant Ri_B \leqslant 0.1$, for $C_{DN} = 1.5 \times 10^{-3}$.

5. The following measurements were made from the research vessel FLIP during the 1969 Barbados Oceanographic and Meteorological Experiment (BOMEX):

   Mean wind speed at a 10.9-m height   = 7.96 m sec^{-1}
   Mean temperature at a 10.9-m height   = 27.34°C
   Mean specific humidity at a 10.9-m height = 16.04 g kg^{-1}
   Sea-surface temperature   = 28.35°C
   Surface pressure   = 1000 mbar

   Using the bulk transfer formulas with $C_D = C_H = C_W = 1.5 \times 10^{-3}$, calculate the surface stress, the sensible heat flux, the rate of evaporation, and the Bowen ratio at the time of these observations.

6. Mean profile measurements made during BOMEX at the same time as

in Problem 5 are given as follows:

Height (m)	Wind speed (m sec^{-1})	Temperature (°C)	Specific-humidity (g kg^{-1})
2.42	7.33	27.55	16.72
3.73	7.58	27.50	16.49
6.18	7.80	27.45	16.29
10.90	7.96	27.34	16.04

(a) Use the profile method with the appropriate Monin–Obukhov flux-profile relations to calculate u_*, θ_*, and q_*.

(b) Compare the fluxes obtained from the profile method with those from the bulk transfer approach.

Chapter 14 | Nonhomogeneous Boundary Layers

14.1 TYPES OF SURFACE INHOMOGENEITIES

Micrometeorological theories, observations, and methods discussed in the preceding chapters are strictly applicable to the atmospheric boundary layer flow over a flat, uniform, and homogeneous terrain. Over such an ideal surface, the PBL is also horizontally homogeneous and in equilibrium with the local surface characteristics, especially when synoptic conditions do not change rapidly. Such a simple state of affairs may exist frequently over open oceans, seas, and large lakes, as well as over extensive desert, ice, snow, prairie, and forest areas of the world. More often, though, land surfaces are characterized by surface inhomogeneities, which make the atmospheric boundary layers over them also nonhomogeneous.

Surface inhomogeneities, which have important effects on atmospheric flows, include boundaries between land and water surfaces (coastlines); the transitions between urban and rural areas or between different types of vegetation; mesoscale oceanic eddies, surface currents, and other regions of varying sea-surface temperature; and hills and valleys. In going over these terrain inhomogeneities, the flow encounters sudden or gradual changes in surface roughness, temperature, wetness, or elevation. Quite often, changes in several surface characteristics occur together, and a complete understanding of modifications in PBL properties (velocity, temperature, and humidity) requires that changes in the surface roughness, temperature, wetness, and elevation be treated simultaneously. For the sake of simplicity and convenience, however, micrometeorologists have studied the effects of changes in the surface roughness, temperature, etc., separately. Due to nonlinearity of the system, however, the individual effects may not simply be added or superimposed; the overall effect of a combination of changes in surface characteristics may differ substantially from the sum of individual effects. There are only a few studies of the combined effects of two or more types of surface inhomogeneities occurring simultaneously.

14.2 STEP CHANGES IN SURFACE ROUGHNESS

Modification of boundary layer following a step change in the surface roughness normal to the direction of flow has been extensively studied experimentally as well as theoretically. A schematic of the approach flow and the modified flow due to change in the surface roughness is shown in Fig. 14.1 for the case of neutral stability. Note that the approach wind-profile $U_1(z)$ is a function of the upwind friction velocity u_{*1} and the surface roughness z_{01}. Following the roughness change from z_{01} to z_{02}, the friction velocity is modified and so is the wind profile near the new surface. Modifications to mean wind profile and turbulence as indicated by computed variances and fluxes are found to be confined to a layer whose thickness increases with distance from the line of discontinuity in the surface. The modified layer is commonly referred to as an internal boundary layer (IBL), because it grows within another boundary layer associated with the approach flow.

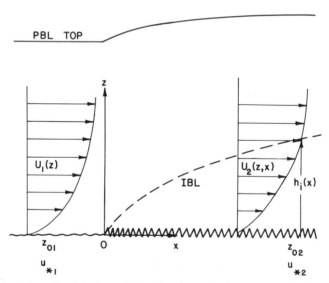

Fig. 14.1 Schematic of the internal boundary layer development and wind profile modification following a step change in the surface roughness.

Above the IBL, flow characteristics are the same as in the approach flow at the same height above the surface. These are essentially determined by upwind surface characteristics. The influence of the upstream surface may be expected to disappear at a sufficiently large distance from the roughness discontinuity, where the IBL has grown to the equilibrium PBL depth for the new surface, or it is otherwise limited by a strong low-level inversion.

In some of the earliest experimental studies of flow over a step change in surface roughness, hundreds of bushel baskets or Christmas trees were placed on a frozen lake (Lake Mendota, Wisconsin) in winter and modifications to the near-surface wind and temperature profiles due to the change in roughness were studied (see, e.g., Kutzback, 1961; Stearns and Lettau, 1963). The observed wind profiles did not extend beyond a height of 3.2 m and no independent shear stress or drag measurements were made in these experiments.

A more comprehensive field study was conducted by Bradley (1968), who measured changes in the surface shear stress and wind profiles with distance from the roughness discontinuity in going from a comparatively "smooth" (but aerodynamically rough) tarmac surface ($z_0 = 0.02$ mm) to a moderately rough surface of wire mesh with spikes ($z_0 = 2.5$ mm), and vice versa. Figure 14.2 shows the variations in surface stress with distance for the two cases (smooth to rough and rough to smooth) of the internal boundary layer flow under near-neutral stability conditions. Here the local stress is normalized by the magnitude of the equilibrium stress for the upwind surface.

Note that the observations of Fig. 14.2 indicate that there is a rather sharp change in the surface stress near the roughness discontinuity, with a considerable amount of overshoot above or undershoot below its expected equilibrium value for the new (rougher or smoother) surface. Relaxation to the equilibrium value occurs very rapidly in the first few meters and gradually thereafter. No detectable changes in the surface stress are apparent from observations beyond a distance of 10–15 m from the roughness discontinuity. But a number of theoretical and numerical model studies predict slow adjustment occurring over much longer distances (see, e.g., Elliot, 1958; Panofsky and Townsend, 1964; Rao et al., 1974); some of these theoretical results are shown in Fig. 14.2.

Observed changes in the wind profile, following step changes in the surface roughness from smooth to rough and vice versa, are shown in Fig. 14.3. At any downwind distance from the surface discontinuity, the lower part of the wind profile is representative of the local surface, while the upper part represents the upwind surface.

Within the lowest 10–15% of the PBL the velocity profile in neutral

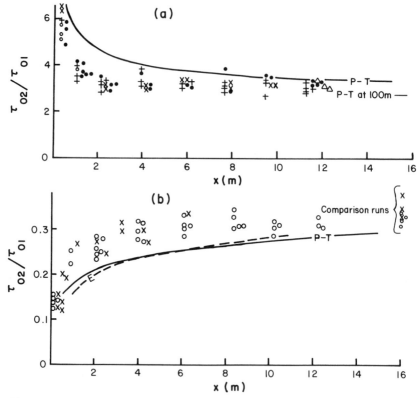

Fig. 14.2 Observed variations of surface shear stress downwind of (a) smooth to rough (from $z_0 = 0.02$ mm to $z_0 = 2.5$ mm) and (b) rough to smooth (from $z_0 = 2.5$ mm to $z_0 = 0.02$ mm) transitions compared with certain theories. ---, Elliot (1958); ——, Panofsky and Townsend (1964). [After Bradley (1968).]

stability may be approximated as (Elliot, 1958)

$$U = (u_{*2}/k) \ln(z/z_{02}), \quad \text{for} \quad z \leq h_i$$
$$U = (u_{*1}/k) \ln(z/z_{01}), \quad \text{for} \quad z > h_i \tag{14.1}$$

In reality, the IBL is not as sharply defined as is implied by the discontinuity (kink) in the velocity profile [Eq. (14.1)], and there is a finite blending region around $z = h_i$ in which the velocity profile gradually changes from one form to another. There is also a question about the validity of the log law in the lower layer, especially in the region where the friction velocity might be changing rapidly and has not reached its equilibrium value u_{*2}. The more appropriate equations of mean motion in the lower part of the

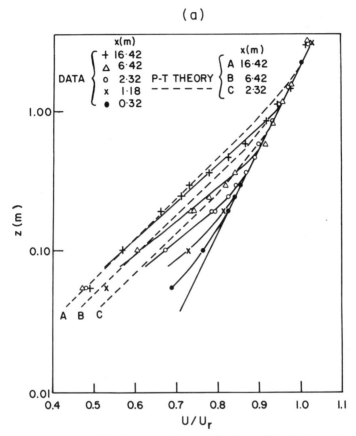

(a)

Fig. 14.3 Observed modifications of velocity profiles downwind of (a) smooth to rough (from $z_0 = 0.02$ mm to $z_0 = 2.5$ mm) and (b) rough to smooth (from $z_0 = 2.5$ mm to $z_0 = 0.02$ mm) transitions compared with Panofsky and Townsend's (1964) theory. [After Bradley (1968).]

IBL, where the Coriolis effects can be neglected, are

$$U(\partial U/\partial x) + W(\partial U/\partial z) = \partial \tau_{zx}/\partial z$$
$$\partial U/\partial x + \partial W/\partial z = 0 \tag{14.2}$$

which form the basis for simpler theoretical models. Note that eliminating the pressure gradient and Coriolis acceleration terms from Eq. (14.2) would limit their validity to rather short distances from the surface discontinuity.

(b)

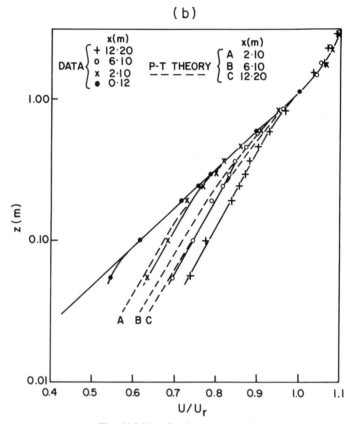

Fig. 14.3(b).   See legend on p. 227.

According to a number of field and laboratory observations taken under neutral stability conditions, the growth of the internal boundary layer thickness follows an approximate power law

$$h_i/z_{02} = a_i(x/z_{02})^{0.8} \qquad (14.3)$$

with estimated values of the empirical constant a_i between 0.35 and 0.75. Actually, a_i depends also on the definition of the IBL thickness. The top of the IBL may be defined as the level where mean wind speed, turbulent momentum flux, or one of the velocity variances reaches a specified fraction (0.90–0.99) of its upstream equilibrium value. An experimental verification of Eq. (14.3) is shown in Fig. 14.4, using Bradley's (1968) field data and those from a wind tunnel simulation. Interestingly, the value of the

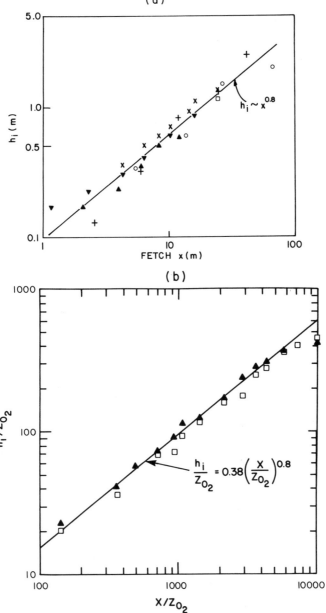

Fig. 14.4 Observed thickness of the internal boundary layer as a function of distance downwind of a step change in the surface roughness based on (a) Bradley's (1968) field data and (b) Pendergrass and Arya's (1984) wind tunnel data. (b) IBL height based on (□) mean velocity and (▲) turbulence. [(b) Reprinted with permission from *Atmospheric Environment*, Vol. 18, W. Pendergrass and S. P. S. Arya, Dispersion in neutral boundary layer over a step change in surface roughness-I. Mean flow and turbulence structure, Copyright (1984), Pergamon Journals Ltd.]

exponent in Eq. (14.3) is the same as that in the approximate classical relationship for the growth of a turbulent boundary layer over a flat plate parallel to the flow (Schlichting, 1960).

Effects of stability on the development of internal boundary layers over warmer and colder than air surfaces have not been systematically investigated. A few experimental and theoretical studies of the IBL under unstable conditions suggest that Eq. (14.3) may also be applicable to these conditions, perhaps with a larger value of a_i than for the neutral case. The IBL growth in a stably stratified atmosphere, on the other hand, may be expected to be slower, suggesting smaller values of the coefficient a_i and possibly also of the exponent in Eq. (14.3). There is a definite need for further studies of the growth and structure of the IBL under stratified conditions.

There have been a few experimental and theoretical studies of the wind-profile modification and the IBL development to large distances of the order of tens of kilometers from the roughness discontinuity, where the Coriolis effects cannot be ignored. It is found that there is a gradual shifting of surface wind direction in a cyclonic or anticyclonic sense in going from smooth to rough or rough to smooth surface, respectively. For example, wind direction shifts of 10–20° have been observed in air flowing from rural to urban areas and vice versa (Oke, 1974). Such changes in the surface wind direction can be explained in terms of the observed dependence of the surface cross-isobar angle (α_0) on surface roughness (α_0 increases with surface roughness, as discussed in Chapter 6). These wind direction changes have important implications for plume trajectories crossing urban and rural boundaries, as well as shorelines. They may also result in distribution of horizontal divergence and convergence and associated distributions of clouds and precipitation.

Many natural surfaces actually present a series of step changes in surface roughness corresponding to different land uses. In principle, one can think of an IBL associated with each step change in surface roughness, so that after a few steps the PBL would be composed of several IBLs. The wind profile is not simply related to the local roughness, except in a shallow surface layer, but represents the integrated effects of several upwind surfaces, depending on the layer (height range) of interest. Observed wind profiles under such conditions often show kinks (discontinuities in slope) when U is plotted against $\log z$; each kink may be associated with an interface between the adjacent IBLs. Such a fine tuning of the wind profile and identification of IBL interfaces may not be possible, however, unless the wind profile is adequately averaged, well resolved, and taken under strictly neutral conditions (most routine wind soundings do not meet these requirements).

14.3 STEP CHANGES IN SURFACE TEMPERATURE

Differences in albedos, emissivities, and other thermal properties of natural surfaces lead to their different surface temperatures, even under a fixed synoptic weather setting. Consequently, inhomogeneities in surface temperature may occur independently of changes in surface roughness. Most dramatic changes in surface temperature occur across lake shorelines and coastlines. In response to these abrupt changes in surface characteristics, thermal internal boundary layers develop over land during onshore flows and over water during offshore flows. Thermal boundary layers also develop in air flow from rural to urban areas and vice versa. On still smaller scales, thermal internal boundary layers develop near the boundaries of different crops, as well as across roads and rivers (see, e.g., Rider *et al.*, 1963).

When temperature differences between two or more adjacent surfaces are large enough and the ambient (geostrophic) flow is weak, thermally induced local or mesoscale circulations are likely to develop on both sides of surface temperature discontinuity. The best known examples of such circulations are the land and sea breezes. Internal boundary layers are often embedded in such thermally driven mesoscale circulations.

14.3.1 THERMAL IBL GROWING OVER A WARMER SURFACE

Perhaps the most dramatic changes in the atmospheric boundary layer occur when stably stratified air flowing over a cold surface encounters a much warmer (relative to air) surface. This frequently occurs in coastal areas during midday and afternoon periods; the resulting flow is called a sea breeze or lake breeze. A similar situation on a large scale occurs during periods of cold-air outbreaks over relatively warm waters (e.g., Great Lakes, Gulf Stream, and Kuroshio) in late fall or winter.

If the temperature difference between downwind and upwind surfaces is fairly large (say, $\Theta_{02} - \Theta_{01} > 5$), the thermal IBL developing over the warmer surface would most likely be very unstable or convective (see Fig. 14.5). The growth of such an IBL can be described by a simple mixed-layer model based on the mean thermodynamic energy equation, under stationary conditions

$$U(\partial\Theta/\partial x) = -(1/\rho c_p)(\partial H/\partial z) \tag{14.4}$$

Integrating Eq. (14.4) with respect to z from 0 to h_i gives

$$h_i U_m(\partial\Theta_m/\partial x) = (1/\rho c_p)(H_0 - H_i) \tag{14.5}$$

in which U_m and Θ_m are the mixed-layer averaged wind and potential temperature and H_0 and H_i are the heat fluxes at the surface and at the top

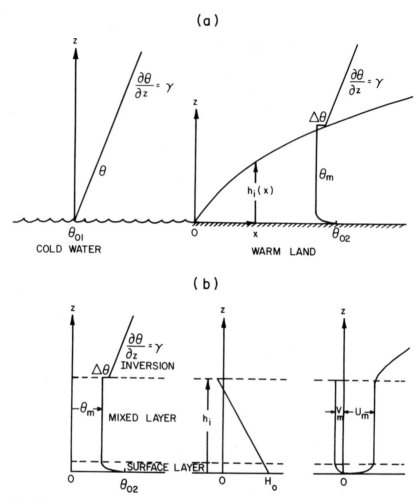

Fig. 14.5 Schematic of a thermal internal boundary layer developing when a stable air flow encounters a much warmer surface and idealized potential temperature, heat flux, and wind profiles in the same.

of the IBL or mixed layer ($z = h_i$), respectively. An equation for the growth of the thermal IBL with distance x from the temperature discontinuity can be obtained after making reasonable assumptions about the potential temperature profile and the heat flux profile, as shown in Fig. 14.5. From these it is clear that

$$\Theta_m = \Theta_{01} + \gamma h_i - \Delta\Theta$$

$$\partial\Theta_m/\partial x = \gamma(\partial h_i/\partial x)$$

(14.6)

where $\gamma \equiv (\partial\Theta/\partial z)_0$ is the potential temperature gradient in the approach flow, which is also the gradient above the thermal IBL. Here, we have assumed that the jump in potential temperature $(\Delta\Theta)$ at the top of the IBL remains constant. If one also assumes that the downward heat flux at the top of the IBL is a constant fraction of the surface heat flux, i.e., $H_i = -AH_0$, Eqs. (14.5) and (14.6) yield an expression for the height of the thermal IBL

$$h_i = a_1(H_0 x/\rho c_p U_m \gamma)^{1/2} \qquad (14.7)$$

in which $a_1 = (2 + 2A)^{1/2} \simeq 1.5$ is an empirical constant. The use of Eq. (14.7) requires the knowledge of the surface heat flux and the mixed-layer wind speed over the downwind heated surface, in addition to that of γ.

An alternative, simpler formula is obtained by expressing or parameterizing the surface heat flux as $H_0 = \alpha\rho c_p U_m(\Theta_{02} - \Theta_{01})$, where $\alpha \simeq 2.1 \times 10^{-3}$ is an empirical coefficient (see, e.g., Vugts and Businger, 1977). Then, Eq. (14.7) can be expressed as

$$h_i = a_2[(\Theta_{02} - \Theta_{01})x/\gamma]^{1/2} \qquad (14.8a)$$

where $a_2 = [2(1 + A)\alpha]^{1/2} \simeq 0.1$. This does not require the knowledge of surface heat flux or mixed-layer wind speed. The latter is introduced in the IBL height expression if one uses the more conventional bulk transfer formula for the surface heat flux

$$H_0 = C_H \rho c_p U_m(\Theta_{02} - \Theta_m)$$

with $C_H \sim C_D = (u_*/U_m)^2$. This type of parameterization leads to the IBL height formula

$$h_i = a_3(u_*/U_m)[(\Theta_{02} - \Theta_m)x/\gamma]^{1/2} \qquad (14.8b)$$

with $a_3 \simeq [2(1 + A)C_H/C_D]^{1/2} \simeq 2.0$. The above expression has the disadvantage of containing the x-dependent variables U_m and Θ_m of the modified mixed layer. Implicit or explicit relations for the variation of Θ_m with x have been given (see, e.g., Fleagle and Businger, 1980). Using some additional assumptions, the IBL height can also be expressed as (Venkatram, 1977)

$$h_i = a_4(u_*/U_m)[(\Theta_{02} - \Theta_{01})x/\gamma]^{1/2} \qquad (14.8c)$$

where $a_4 \simeq 1.7$. Some of the above expressions for h_i have been compared and verified against observations (see, e.g., Raynor et al., 1979; Stunder and SethuRaman, 1985). Equation (14.7) appears to be the best, because its derivation involves the least number of assumptions.

The above-mentioned formulations of the IBL growth are based on mixed-layer assumptions for the IBL and neglect of the near-surface layer, in which velocity and potential temperature generally have strong gradients. These are not expected to be valid for short (less than 1 km) distances from the temperature discontinuity, where modifications are likely to be confined to the shallow surface layer. Detailed experimental investigations of air modification in the surface layer in the immediate vicinity of a step change in surface temperature have been reported by Rider et al. (1963) and Vugts and Businger (1977). Here, the growth of the thermal IBL is more rapid and more closely follows the $x^{0.8}$ behavior, similar to that of an IBL following a step change in roughness (see, e.g., Elliot, 1958).

14.3.2   THERMAL IBL GROWING OVER A COLDER SURFACE

As compared to the rapidly growing and highly energetic unstable or convective thermal IBL developing in cold air advecting over a much warmer surface, the reverse situation of comparatively shallow and smooth stable IBL that develops in warm air advection over a colder surface has received very little attention. An early observational study was made by Taylor (1915) on the ice-scout ship S.S. Scotia near the coast of Newfoundland. More recently, Raynor et al. (1975) have reported on some measurements of the depth of the stable IBL at the southern shore of Long Island, New York, as a function of fetch. The IBL resulted from modification of warm continental air following a south westerly trajectory over colder oceanic waters; its depth could be represented by an empirical relationship

$$h_i = a_5(u_*/U)[(\Theta_{01} - \Theta_{02})x/|\partial T/\partial z|]^{1/2} \qquad (14.8d)$$

which is similar to Eq. (14.8c) for the unstable IBL. Here, $|\partial T/\partial z|$ is the absolute value of the lapse rate or temperature gradient over the source region or above the inversion in the modified air, U is the mean wind speed at a reference level (say, 10 m) near the surface, and a_5 is an empirical constant of the order of unity which may depend on the choice of the reference level for measuring or specifying wind speed.

Note that despite the apparent similarities of the IBL height equations [Eqs. (14.8c) and (14.8d)] in different situations of cold- and warm-air advection, respectively, for a given fetch the magnitudes of h_i and $\partial h_i/\partial x$ are found to be much smaller in the latter case. These and other IBL height equations showing unrestricted growth of internal boundary layers with fetch or distance from the discontinuity in the surface roughness, temperature, etc., must cease to be valid beyond a certain distance where the IBL has grown to the equilibrium depth of the PBL for the given surface and external conditions. Observations indicate that this might take tens to hundreds of kilometers, depending on the equilibrium PBL depth.

14.4 AIR MODIFICATIONS OVER WATER SURFACES

When continental air flows over large bodies of water, such as large lakes, bays, sounds, seas, and ocean, it usually encounters dramatic changes in surface roughness and temperature. Consequently, significant modifications in air temperature, specific humidity, cloudiness, winds, and turbulence occur in the developing IBLs with distance from the shoreline. Most dramatic changes in air properties and ensuing weather phenomena occur when cold air flows over a much warmer water surface. Intense heat and water vapor exchanges between the water surface and the atmosphere lead to vigorous convection, formation of clouds, and, sometimes, precipitation. Some examples of this phenomenon are frequent cold-air outbreaks over the warmer Great Lakes in late fall and early winter and the resulting precipitation (often, heavy snow) over immediate downwind locations. It has been the focus of several field experiments, including the International Field Year of the Great Lakes (IFYGL). Figure 14.6 presents the observed changes in potential temperature and cloudiness across Lake Michigan during a cold-air outbreak when the water–air temperature difference was about 12.5 K (Lenschow, 1973). Note the rapid warming and moistening of the approaching cold, dry air mass by the warmer lake surface water.

Other major investigations of air modification during cold-air outbreaks over warm waters have been conducted during the 1974 and 1975 Air Mass Transformation Experiments over the Sea of Japan and a more recent (1986) Genesis of Atlantic Lows Experiment (GALE). In the latter experiment was observed one of the most intense cold-air outbreaks with the sea–air temperature differences of 20–25 K over the Gulf Stream just off the North Carolina coast. The ocean-surface temperature here varies considerably with distance from the coast, with several step changes between the coast and the eastern edge of the Gulf Stream. Several research aircraft, ships, and buoys were used to study air-mass modifications in the boundary layer at different locations. The cold-air outbreak of January 28, 1986, was most spectacular in that the ocean surface was enshrouded in a steamlike fog with abundant number of steam devils, water spouts, and other vortexlike filaments visible from the low-flying aircraft. The PBL depth increased from about 900 m at 35 km offshore to 2300 m at the eastern edge of the Gulf Stream (284 km offshore). At the same time, the cloudiness increased from zero to a completely overcast,

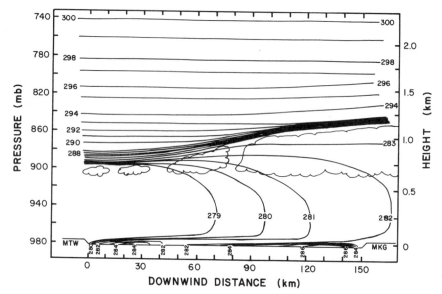

Fig. 14.6 Observed modifications in potential temperature and cloudiness across Lake Michigan during a cold-air outbreak (November 5, 1970). Mean wind is from left to right. [After Lenschow (1973).]

thick deck of stratocumulus with snow and rainshowers developing over the eastern edge of the Gulf Stream. The subcloud mixed layer was characterized by intense convective turbulence with strong updrafts and downdrafts.

A simple theory of predicting changes in air temperature and specific humidity as a function of fetch for the case of cold-air advection over a warm sea, but in the absence of condensation and precipitation processes in the modified layer, is given by Fleagle and Businger (1980). When the top of the IBL reaches above the lifting condensation level, however, the structure of the IBL changes dramatically. Complex interactions take place between cloud particles, radiation, and entrainment at the top of the IBL.

Modifications of warm continental air advecting over a cold lake or sea surface are equally dramatic, but the modified layer is much shallower. Turbulence is considerably reduced in this shallow, stably stratified IBL, while visibility may be reduced due to fog formation under appropriate conditions. Cases of strong warm-air advection commonly occur in springtime over the Great Lakes and along the east coast of the United States and Canada. Figure 14.7 presents some observations of air modification under such conditions prevailing over Lake Michigan (Wylie and

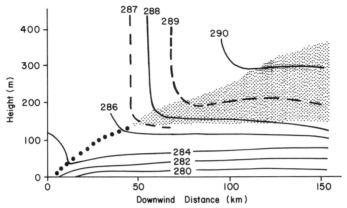

Fig. 14.7   Observed modifications in potential temperature over Lake Michigan during warm-air advection. Surface-based inversion is unshaded area below the dotted line; shading denotes stable isothermal layer. [After Wylie and Young. Copyright © (1979) by D. Reidel Publishing Company. Reprinted by permission.]

Young, 1979). These were made from a ship traveling in the downwind direction across the lake, with instruments attached to a tethered balloon. Note that the surface-based inversion that formed over water rose to about 140 m after a fetch of 50 km and remained at a constant level for the rest of the fetch along the lake. A growing isothermal layer separated the surface inversion layer from the adiabatic mixed layer advected from upwind land areas. Large variations in wind speed and direction were observed within the inversion layer. Note that the thermal internal boundary layer, including the isothermal layer, kept growing over the whole fetch of more than 150 km across the lake.

14.5   AIR MODIFICATIONS OVER URBAN AREAS

Urbanization, which includes residential, commercial, and industrial developments, produces radical changes in radiative, thermodynamic, and aerodynamic characteristics of the surface from those of the surrounding rural areas. Therefore, it is not surprising that as urbanization proceeds the associated weather and climate often are modified substantially. Such modifications are largely confined to the so-called urban boundary layer, although the urban "plume" of pollutants originating from the city may extend hundreds of kilometers downwind. Many studies of urban influences have identified significant changes in surface and air temperatures, humidity, precipitation, fog, visibility, air quality, surface energy fluxes, mixed-layer height, boundary layer winds, and turbulence between the urban and rural areas. Here, we give a brief description of some of the better known urban-induced phenomena.

14.5.1 THE URBAN HEAT ISLAND

The most frequently observed (or felt) and best documented climatic effect of urbanization is the increase in surface and air temperatures over the urban area, as compared to rural surroundings. Closed isotherms over the urban and suburban areas generally separate these areas from the rural environs. This condition or phenomenon has come to be known as the urban heat island, because of the apparent similarity of isotherms to contours of elevation for a small, isolated island in the ocean. This analogy is carried further in the schematic representation of near-surface temperature in Fig. 14.8 for a large city on a clear and calm evening, as one travels from the countryside to the city center. This shows a typical "cliff" of steep rise in temperature near the rural/suburban boundary, followed by a "plateau" over much of the suburban area, and then a "peak" over the city center (Oke, 1978). The maximum difference in the urban peak temperature and the background rural temperature defines the urban heat island intensity (ΔT_{u-r}). Over large metropolitan areas, there may be several plateaus and peaks in the surface temperature trace. Also, there are likely to be many small scale variations in response to distinct intraurban land uses, such as parks, recreation areas, and commercial and industrial developments, as well as topographical features, such as lakes, rivers, and hills. Strictly urban influences on the local weather and climate can be studied in an isolated manner only in certain interior (as opposed to coastal) cities in relatively flat terrain. For this reason, St. Louis, Missouri, was chosen to be the arena for several large field studies, viz., the Metropolitan Meteorological Experiment (METROMEX, 1971–1976) and Regional Air Pollution Study (RAPS, 1973–1977), of urban influences on local weather and climate.

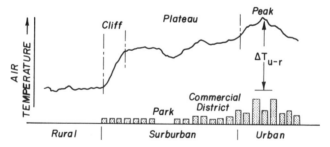

Fig. 14.8 Schematic representation of variation in air temperature in going from a rural to an urban area. [After Oke (1987).]

Figure 14.9 shows a map of the St. Louis metropolitan area representing major land-use types. The principal urban area is located about 16 km south of the confluence of the Mississippi and Missouri rivers. The primary topographic features in the area are the river valley with a prominent bluff rising 20–60 m above the flood plain east of the merged river and heavily wooded (70–90 m in height) hills to the southwest. A slightly smoothed pattern of aircraft-observed (with an infrared radiometer) ground-surface temperatures on a clear, summer afternoon is shown in Fig. 14.10. Note that it clearly shows the urban heat-island phenomenon with an intensity of $\Delta T_{u-r} \simeq 8.4$ K at that particular time. The difference in air temperatures at a 10-m height is likely to be much smaller.

Other observations of an urban heat island over small and large cities have been reviewed by Landsberg (1981) and Oke (1974). The intensity of an urban heat island depends on many factors, such as the size of city and its energy consumption, geographical location, month or season, time of day, and synoptic weather conditions. The maximum intensity for a given city occurs in clear and calm conditions (e.g., under a stationary high), a few hours after sunset. The intensity becomes minimal or zero under highly disturbed (stormy), windy weather conditions.

Under reasonably stationary synoptic weather conditions, the heat-island intensity shows a pronounced diurnal variation with a minimum value around midday and maximum value around or before midnight (see, e.g., Oke, 1978, Chap. 8). There are likely to be marked differences in the diurnal variations of ΔT_{u-r} between winter and summer related to the anthropogenic heat release.

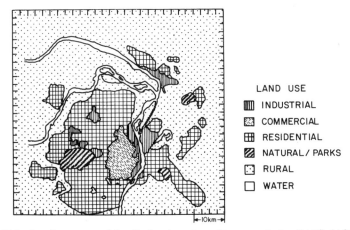

LAND USE
▦ INDUSTRIAL
▨ COMMERCIAL
⊞ RESIDENTIAL
▨ NATURAL / PARKS
⦂ RURAL
▢ WATER

←10km→

Fig. 14.9 Land use map of the St. Louis metropolitan area during RAPS. [After Byun (1987).]

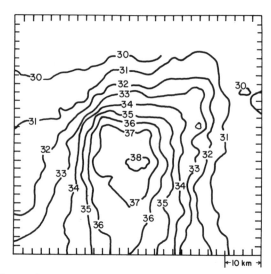

Fig. 14.10 Observed patterns of ground-surface temperatures showing the heat-island phenomenon over the St. Louis metropolitan area on a clear summer afternoon during RAPS. [After Byun (1987).]

Some attempts have been made to correlate the maximum nocturnal heat-island intensity (based on air temperatures at 10 m above ground level) with population (considered as a surrogate for the city size and energy consumption) and near-surface wind speed. Figure 14.11 shows different relationships of $\Delta T_{(u-r)max}$ with population (P) for cities of Europe and North America with minimal topographical influences. Different curves probably reflect differences in the per capita energy consumption and in population density between European and North American cities. The data and regression lines in Fig. 14.11 are for calm and cloudless conditions. For moderate wind speeds, ΔT_{u-r} is found to decrease in inverse proportion to the square root of regional (nonurban) wind speed, and vanishes at a "critical" wind speed of 10 m sec^{-1} (measured at a height of 10 m at a rural site) for large cities (Oke, 1978, Chap. 8).

The extent and intensity of the urban heat island is also greatly influenced by topography and the presence of large bodies of water within or around the urban area. In coastal cities, land and sea breezes have a profound influence, while in hilly areas nocturnal drainage and valley flows may dominate over the strictly urban influences.

What are the causes for the urban heat-island phenomenon? A number of causes have been hypothesized and most of them are verified by observations. The leading candidates are (Oke, 1978, Chapter 8) as follows:

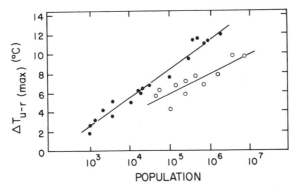

Fig. 14.11 Empirical relation between the observed heat-island intensity and city population for (●) North American and (○) European cities. [After Oke (1987).]

- Increased incoming longwave radiation ($R_{L\downarrow}$) due to absorption of outgoing longwave radiation and reemission by polluted urban atmosphere
- Decreased outgoing longwave radiation loss ($R_{L\uparrow}$) from street canyons due to a reduction in their sky view factor by buildings
- Increased shortwave radiation (R_S) absorption by the urban canopy due to the effect of street canyons on albedo
- Greater daytime heat storage (ΔH_s) due to the thermal properties of urban materials and heat release at nighttime
- Addition of anthropogenic heat (H_a) in the urban area in process emission (heating and cooling), transportation, and industrial operations
- Decreased evaporation and, hence the latent heat flux (H_L) due to the removal of vegetation and surface waterproofing of the city

In short, all the components of the energy balance in the urban canopy are modified in such a way that they add to the heat-island effect in the positive sense. Some of these are effective only during daytime when they contribute to the storage of heat in the urban canopy; heat release at nighttime keeps the urban air warmer. The relative role of each component is likely to differ from one season to another and also from one city to another.

14.5.2   THE URBAN BOUNDARY LAYER

The urban boundary layer (UBL) is modified by both the urban heat island and the increased roughness (one can also think of an urban area as a "roughness island"). In the absence of any topography, the roughness elements of a city are its buildings, whose heights (h_0) generally increase toward the city center. Practical considerations make it difficult to determine the value of the roughness length or parameter (z_0) directly from wind-profile measurements over a large city center or commercial area; such observations should be made above the prevailing height of the buildings. But rough estimates can be made by extrapolating the z_0/h_0 and d_0/h_0 relationships discussed in Chapter 10 (as in the case of tall vegetation, the roughness parameterization of an urban canopy must include a zero-plane displacement). In an urban area, the roughness parameter may vary from fractions of a meter in suburbs to several meters over the city center.

In near-calm or weak-wind and clear-sky conditions, thermal modification of the boundary layer in response to the urban heat island is likely to dominate over increased roughness effects. The thermally induced circulation that is superimposed on any weak background flow is radially inward toward the city center at lower levels and outward from the city center at upper levels, with a rising motion over the center and susidence motion over the surrounding environs. During the day this circulation may extend up to the base of the lowest inversion, which may then acquire a dome shape in response to the induced circulation (rising and subsiding motions). Since mixing is likely to be restricted to the mixed layer below the inversion, it is also referred to as the ''mixed-layer dome,'' or ''dust dome,'' in which dust, smoke, and haze from urban emissions accumulate during stagnant conditions. Figure 14.12 depicts a schematic of thermally induced circulation and dust dome over an urban area. Figure 14.13 shows the same (mixed-layer dome) with aircraft-observed profiles of potential temperature and specific humidity in the UBL of St. Louis, Missouri (Spangler and Dirks, 1974). At that time, winds were nearly uniform in the lowest 500 m, variable above this height up to the inversion base, and again uniform at 5 m sec^{-1} above the inver-

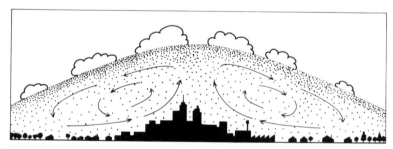

Fig. 14.12 Schematic of dust dome and thermally induced circulations over a large city under calm or light winds. [From ''The Climate of Cities'' by W. P. Lowry. Copyright © (1967) by Scientific American, Inc. All rights reserved.]

sion. Note that a rural-to-urban variation in mixed-layer height of at least 400 m was observed even when the urban heat-island intensity was not particularly strong $T_{u-r} < 1$ K).

Aircraft measurements of turbulent fluxes and variances in the daytime UBL also indicate considerable urban-scale variations of these parameters (see, e.g., Ching, 1985). For example, observations over St. Louis show that the sensible heat flux varied by a factor of 2 to 4, with the largest values over the city center. The latent heat flux also varied by a factor of 4, but with smallest values over the city center. Consequently, the Bowen ratio exhibited even larger spatial variability, with a maximum value of about 1.8 over the city center and a minimum value of less than 0.2 over some nonurban areas. Spatial patterns of turbulent velocity variances are similar to those of the sensible heat flux.

At night the urban boundary layer shrinks to a depth of a few hundred meters because strong stability of the approach flow suppresses turbulent mixing in the vertical direction. Still, the UBL over the city center is much thicker than the nocturnal stable boundary layer upwind of the city. The combination of increased temperatures (heat island) and increased roughness over a moderate-size and a large-size city can easily destroy the nocturnal surface inversion and dramatically modify the nocturnal boundary layer as it advects over the city. This process or phenomenon is illustrated in Fig. 14.14, which shows a helicopter-measured along-wind vertical temperature cross-section, as well as the vertical potential temperature profiles at different locations in Montreal, Canada, on a winter

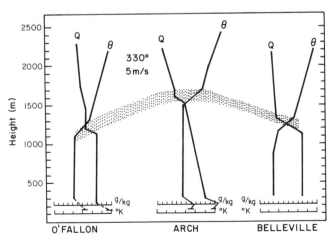

Fig. 14.13 Observed mixed-layer dome and potential temperature and specific humidity profiles over St. Louis on August 12, 1971. [After Spangler and Dirks. Copyright © (1974) by D. Reidel Publishing Company. Reprinted by permission.]

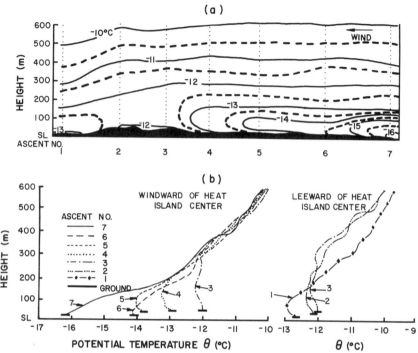

Fig. 14.14 Modifications of potential temperature in the UBL due to the urban heat island over Montreal on a winter morning: (a) x–z cross-sectional distribution of Θ and (b) vertical profiles of Θ at the various locations along wind. [After Oke and East. Copyright © (1971) by D. Reidel Publishing Company. Reprinted by permission.]

morning. Other measurements of air temperature from automobile traverses in the city indicated an urban heat-island intensity of about 4.5 K at that time (Oke and East, 1971).

Figure 14.14 shows that air is progressively warmed as it traverses the urban area (the weak flow from the N/NE was along the main helicopter traverse route), and the depth of the modified urban boundary layer increases downwind. At the rural site (sounding location 7) the atmosphere is very stable from the surface up to at least 600 m. In moving over the urban area, the bottom (urban canopy) layer becomes unstable and the layers above become neutral or weakly stable. Above the top of the UBL, the capping inversion layer retains the characteristics of the "rural" approach flow. The UBL attains its maximum depth (300 m) over the city center and appears to become shallower farther downwind of the city center. A surface inversion may also reform over the downwind rural area. The modified urban air advecting above this shallow surface inversion layer is called the urban plume.

In comparison to the strong diurnal changes in stability of the lower atmosphere in surrounding rural areas, the urban atmosphere experiences only small diurnal variations in stability. The UBL remains well mixed both by day and by night, although the mixing depth usually undergoes a large diurnal oscillation. The difference in the stabilities of the rural and urban boundary layers at nighttime explains why surface winds are often greater in the latter. Strong inversion reduces the near-surface winds in the rural area and essentially decouples them from stronger winds aloft. More efficient vertical mixing in the UBL, on the other hand, results in the increased momentum (winds) in the surface layer.

Observational studies of the UBL have shown that the urban heat-island phenomenon, quantified by the urban–rural temperature difference, extends through the whole depth of the UBL. However, the heat-island intensity is maximum at the surface, decreases with height, and vanishes at the top of the UBL. There is also some evidence of the so-called cross-over effect, according to which the heat-island intensity becomes slightly negative above the top of the UBL (Oke and East, 1971). This effect is observed only at low wind speeds ($U < 3$ m sec^{-1}).

A simple but appropriate relationship between the intensity of a nocturnal urban heat island and the maximum mixing depth over a city is

$$h_u = (\Theta_u - \Theta_r)/(\partial\Theta_r/\partial z) \simeq (T_u - T_r)/[(\partial T_r/\partial z) + \Gamma] \qquad (14.9)$$

in which T_u and T_r are urban and rural air temperatures near the surface. It is based on the assumptions that the rural temperature sounding can be characterized by a constant gradient, at least above the shallow surface inversion layer, and there is no large-scale cold- or warm-air advection. Good agreement has been found between the predicted and observed values of h over New York, Montreal, and other cities.

14.6 BUILDING WAKES AND STREET CANYON EFFECTS

In the preceding section we discussed the urban effects on atmospheric boundary layer well above the urban canopy, and details of three-dimensional, complex flows around buildings were glossed over. For urban inhabitants, the local environment on streets around buildings, i.e., within the urban canopy, is perhaps more important than that in the above-canopy boundary layer. The canopy flow is much more complex, however, and is not amenable to any simple, generalized, quantitative treatment. Therefore, we give here only a brief qualitative description of some of the observed features of flow around buildings and other urban structures. These observations have largely come from physical simulations of flow around isolated and somewhat idealized buildings, as well as from clusters of buildings in environmental wind tunnels. Most of our conceptual understanding of such flows is derived from fundamental studies in bluff-body aerodynamics and fluid mechanics.

14.6.1 CHARACTERISTIC FLOW ZONES AROUND AN ISOLATED BUILDING

There have been many experimental studies of flow distortion (modification) around isolated bluff bodies (circular and rectangular cyclinders, flat plates, etc.) placed in a uniform, streamlined (laminar) approach flow. Comparatively fewer studies have been made of the distorted flow fields around surface-mounted structures immersed in thick boundary layers, as in the atmosphere. Although details of the modified flow vary with the characteristics of the approach flow, as well as with the size, shape, and orientation of the bluff body, certain flow phenomena and characteristic zones are commonly observed.

14.6.1.1 Flow Separation and Recirculating Cavity

Flow separation is said to occur when fluid initially moving parallel to a solid surface suddenly leaves the surface, because it can no longer follow the surface curvature or because of a break (discontinuity) in the surface slope. Flow separation is essentially a viscous flow phenomenon, as an inviscid or ideal fluid can, in principle, go around any bluff body without separation. The surface boundary layer developing in a real fluid flow, on the other hand, usually separates and moves out into the flow field as a free shear layer. The flow near the surface immediately downstream of the boundary layer separation is in the opposite direction. This is illustrated by the schematics of flow separation from a curved surface (Fig. 14.15a) and a sharp-edged surface (Fig. 14.15b). Note that the surface streamline leaves off tangentially or parallel to the initial slope from the separation point S.

The primary cause of flow separation is the loss of mean kinetic energy (e.g., in a diverging flow) in the boundary layer and consequent gain in the potential energy and, hence, in surface pressure in the direction of flow, until the flow is no longer possible against the increasing adverse pressure gradient. All fluids have a natural tendency to flow in the direction of decreasing pressure, i.e., along a favorable pressure gradient. A significant adverse (positive) pressure gradient in the direction of flow can lead to abrupt and dramatic changes in the flow following its separation from the boundary.

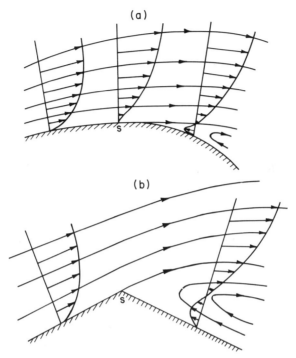

Fig. 14.15 Schematic of flow separation from a (a) curved surface and (b) sharp-edged surface, and typical velocity profiles and streamline patterns.

The flow around a curved surface, as shown in Fig. 14.15a, often leads to an unsteady separation in which the separation point S may not remain fixed but moves up and down the surface in response to perturbations to the flow, including the shedding of vortices by the separated shear layer. Even the average location of S is found to be quite sensitive to slight surface irregularities, as well as to the Reynolds number (Re), based on the radius of curvature of the surface. On the other hand, the flow separation at a salient edge, as shown in Fig. 14.15b, is steady, and its location is fixed, irrespective of the Reynolds number. This property is found to be quite useful in reduced-scale model simulations of atmospheric flows around buildings in fluid-modeling facilities, such as wind tunnels and water channels, in which Reynolds numbers are necessarily much smaller than those in the atmosphere. Gross features of flow around sharp-edged bluff bodies are found to be essentially Reynolds-number independent for all large enough Reynolds numbers (say, Re $> 10^4$).

An example of flow separation from the upwind edge of the roof of a

(a)

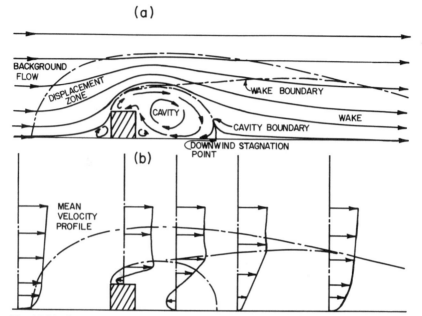

Fig. 14.16 Schematic of displacement, cavity, and wake flow zones around a two-dimensional sharp-edged building: (a) mean streamline pattern and (b) mean velocity profiles at the various locations along the flow. [After Halitsky (1968).]

rectangular building is shown in Fig. 14.16, in which other flow zones are also depicted. Note that the separated streamline eventually reattaches to the ground (or the roof in the case of a long building) at the downwind stagnation point. Below this a relatively stagnant zone of recirculating flow, or "cavity," is formed. Contaminants released into this region or those entrained from above often lead to very high concentrations in the cavity region, inspite of some cross-streamline diffusion and mixing with the outer fluid. The cavity region is characterized by a recirculating mean flow with low speeds but large wind shears and high-turbulence intensities. Flow separation also occurs from the sides of the building (not shown in Fig. 14.16), so that the three-dimensional cavity is a complex structure.

Dimensions of the cavity envelope depend on the building configuration (aspect) ratios (W/H and L/H, where L, W, and H are the building length, width, and height, respectively), as well as on the characteristics of the approach flow, such as the boundary layer depth relative to the building height (h/H). The maximum cavity length may vary from H (for $W/H < 1$) to $15H$ (for $W/H > 100$, and $L/H < 1$), while the maximum depth of the cavity can vary between H and $2.5H$. For a more comprehensive review of wind tunnel and field data on cavity dimensions and empirical relations based on the same, the reader should refer to Hosker (1984).

Figure 14.16 also shows a small frontal cavity or vortex region confined to the upwind ground-level corner of the building. This is caused by the separation of the flow from the ground surface, resulting from an adverse pressure gradient immediately upwind of the building (note that the pressure is maximum at the stagnation point on the upwind face of the building).

14.6.1.2 Wake Formation and Relaxation

The region of flow immediately surrounding and following the main recirculating cavity, but still affected by the building, is called the building "wake." According to some broader definitions, the wake also includes the lee-side cavity region, sometimes called the "wake cavity," and comprises the entire downstream region of the flow affected by the obstructing body (all bluff and streamline bodies have wakes, even though some may not have any recirculating cavity in their lee). Here we use the more restrictive definition excluding the cavity from the wake, so that there is no flow reversal in the latter.

Sometimes the wake is subdivided into the "near wake" and the "far wake." The mean flow and turbulence structure in the former is strongly affected by separating shear layers, as well as by vortices shed from the building edges, as shown in Fig. 14.17. The separated shear layers have large amounts of vorticity, generated when these layers were still attached to the building as boundary layer. They also have strong wind shears across them. After separation the shear layers become unstable and generate a lot of turbulence and, frequently, also periodic vortices called Karman vortices, which grow as they travel downwind. Another prominent vortex system in flows around three-dimensional surface obstacles is the so-called horseshoe vortex. It is a standing horizontally oriented vortex generated near the ground upwind of the obstacle; it wraps around the obstacle and then trails off downwind as a counterrotating vortex pair (see Fig. 14.17). This vortex resembles a horseshoe when viewed from above, hence its name. These large vortices produce oscillations in the wake boundary, while smaller eddies transport momentum across mean streamlines. As a consequence of these processes, mean velocity in the direction of flow decreases rapidly, while turbulence increases with downwind distances in the near wake, which may extend up to a few building heights from the upwind face of the building. Building-

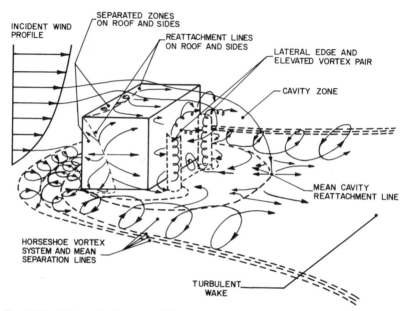

INCIDENT WIND
PROFILE

SEPARATED ZONES
ON ROOF AND SIDES

REATTACHMENT LINES
ON ROOF AND SIDES

LATERAL EDGE AND
ELEVATED VORTEX PAIR

CAVITY ZONE

MEAN CAVITY
REATTACHMENT LINE

HORSESHOE VORTEX
SYSTEM AND MEAN
SEPARATION LINES

TURBULENT
WAKE

Fig. 14.17   Schematic of separated flow zones and vortex systems around a three-dimensional sharp-edged building in a deep boundary layer. [After Woo *et al.* (1977).]

generated turbulence is found to be predominant in the near wake and cavity regions.

The near wake is followed by a more extensive far-wake region in which perturbations to mean flow and turbulence decay with distance, following an exponential or power law, and become insignificant only at large downwind distances. The longitudinal, lateral, and vertical extents of the wake depend on building dimensions and their aspect ratios. Both the width and the thickness (height) of the wake grow with distance behind the building, if the wake is defined as the region encompassing all building-induced perturbations to the background or approach flow, even if they become practically undetectable. This is the classical theoretical definition of wake, which extends to infinity along the flow direction as it grows in cross-wind directions. A more practical definition of the wake in which perturbations in the mean and turbulent flow are significant, i.e., they are larger than a certain specified percentage (say, 10%) of the background flow, would make the wake boundary a finite envelope with a maximum length, width, and height, which depend on building aspect ratios. Wind tunnel observations in simulated building wakes indicate that significant building-induced perturbations to flow may not extend beyond 10 to 20 building heights in the downwind direction and 2 to 3 building heights or widths in other directions. Sometimes, the far wake is further subdivided into an inner layer or a surface layer, and an outer mixing layer. The former is assumed developing downwind of the cavity reattachment point and has the character of an internal boundary layer in which flow adjusts to the local surface (see Plate, 1971, Chap. 4). The outer mixing layer has the character of a classical wake far from any bounding surface.

14.6.1.3   External Region of Streamline Displacement

A third region of relatively minor modifications to the flow also envelops the building and its wake and cavity regions. Streamlines in this so-called external or displacement region are at first deflected upward and laterally outside in response to the growing wake, and then downward and inside to their undisturbed pattern, as the building influence decays and disappears. These displacements and associated perturbations to the flow decrease with increasing distance outward from the wake boundary. Such perturbations to the flow in the external region are considered to be essentially inviscid and small, allowing for simplified theoretical treatments. Note that external zones of streamline displacement must also exist, to some extent, in all the nonhomogeneous boundary layers (see, e.g., Townsend, 1965). Outside the displacement zone is the region of undisturbed flow.

14.6.2   CLUSTERING AND STREET CANYON EFFECTS

In urban settings, buildings are placed in clusters of individual houses, high-rise apartments, or commercial buildings. When the spacing between adjacent buildings is less than 10 to 20 building heights, which is generally the case, the wakes and cavities associated with individual buildings interact, producing a variety of complicated and often discomforting flow patterns. There have been only a few simulation studies of clustering effects of buildings in urban settings, and no field study at all (see, e.g., Hosker, 1984). Theory and numerical simulations cannot cope with such complex and highly turbulent flows. Experimental efforts are also beset by serious instrumental and logistical difficulties. Our qualitative understanding of flow phenomena associated with building clusters is largely due to wind tunnel simulation studies.

When buildings are nearly of the same size and height, and are arranged in regular rows and columns parallel to straight streets in an otherwise flat area, air flow is accelerated just above the roof level outside of any roof cavities, as well as in any side streets parallel to the wind direction. Winds can become strong and steady, particularly in narrow and deep street canyons, due to direct blocking of flow by windward faces of the buildings and channeling in side-street canyons. Streets normal to the ambient wind direction are relatively sheltered when the ambient atmospheric boundary layer flow is weak, but strong standing vortices can develop in cross streets at moderate ambient wind speeds. These are augmented forms of lee-side cavities where the recirculating flow is enhanced by deflection down the windward face of the adjacent downstream building. Winds are more gusty (turbulent) in the presence of these building-induced vortices. If the ambient wind flow is at an angle to the streets or buildings, lee-side vortices take on a "corkscrew" motion with some along-street movement. The winds in the side streets are also reduced and become more gusty, reflecting intensification of corner vortices. All of the above-mentioned flow patterns strongly depend on building aspect ratios and the relative spacing between them. Irregular arrangements of building blocks along curved streets can result into an almost infinite variety of flow patterns. Some of the basic fluid flow phenomena underlying these complex flow patterns are schematically illustrated and discussed by Hosker (1984).

When adjacent buildings in a cluster differ considerably in their heights or aspect ratios, flow patterns around them become even more complex and highly asymmetrical. Although a wide variety of situations can be visualized and actually occur in urban settings, systematic experimental studies (primarily wind tunnel simulations) are so far confined to the so-called two-body problem in which a much taller building is placed downwind of a shorter building, or vice versa (see, e.g., Hosker, 1984). Here, too, the flow patterns depend strongly on the relative building heights, spacing between the two buildings, and the building aspect ratios, as well as on the characteristics of the approach flow. The presence of a much smaller building upwind of a tall building appears to affect only the upwind portion of the horseshoe vortex, which wraps around any isolated building, and the downwind cavity and wake of the taller building are less affected. These can be considerably affected, however, if the shorter building is placed downstream of the taller building. The taller building has a more dominating influence on the circulation around a smaller building in its wake than the other way around.

14.7 OTHER TOPOGRAPHICAL EFFECTS

In this section we consider the effects of small hills, ridges, and escarpments on atmospheric boundary layer flows. Effects of large mountains extend well above the boundary layer and involve mesoscale and, sometimes, synoptic-scale motions, which are outside the scope of this book. Unlike the flows around sharp-edged buildings or groups of buildings in which mechanically generated turbulence dominates over the ambient turbulence in the approach flow, flows around gentle, low hills and ridges are quite sensitive to mean shear and turbulence in the approach flow. Therefore, their description is presented here according to different approach flow conditions. This is largely based on laboratory fluid-modeling experiments, some theoretical and numerical model studies, and a few field experiments of recent years. For a more comprehensive review of the literature, the reader should refer to Hunt and Simpson (1982).

14.7.1   NEUTRAL BOUNDARY LAYER APPROACH FLOW

The characteristic flow zones shown in Fig. 14.15 for the boundary layer modified by an isolated building can also be applied to neutral boundary layer flow around an isolated hill, ridge, or escarpment. However, flow separation and associated cavity regions are likely to be unsteady and much smaller, if they exist at all, around gentle topographical features. The flow may not separate even on the lee side, unless the maximum downwind slope is large enough (say, ≥ 0.2). Behind long, steep ridges, however, the cavity region may extend up to 10 hill heights in the downstream direction. The cavity is at most a few hill heights long and frequently smaller for three-dimensional hills.

The lee-side hill wakes are similar to building wakes; these are characterized by reduced mean flow and enhanced turbulence. The maximum topographically induced perturbations to the flow in the near wake depend on the aspect ratio, slope, and shape of the hill. However, their decay with distance in the far wake appears to follow similar relations. In particular, the maximum velocity deficit $(\Delta U)_{max}$ is given by

$$(\Delta U)_{max}/U_0(H) = m(x/H)^{-1} \tag{14.10}$$

where $U_0(H)$ is the mean velocity at hill height in the approach flow, and m is a coefficient depending on the hill and boundary layer parameters. The rate of growth of the wake also depends on the hill slope and aspect ratio (see, e.g., Arya et al., 1987).

An important universal feature of the flow over a hill is its speeding up in going over the hill's top. For quantifying this, a speed-up factor $S_F = U(z)/U_0(z)$ is defined as the ratio of wind speed at some height above the hill to that at the same height above the flat surface in the approach flow. While S_F decreases with increasing height, its maximum values range from just over unity to about 2.5, depending on the hill shape, slope, and

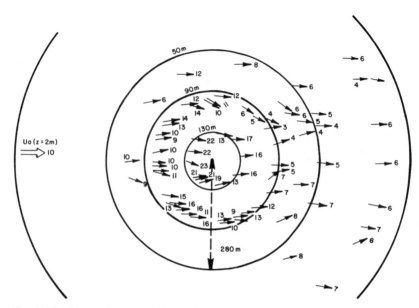

Fig. 14.18 Observed mean wind speeds and directions near the surface ($z = 2$ m) of an approximately circular low hill (Brent knoll). [After Mason and Sykes (1979).]

aspect ratio. The largest speed-up factors are observed over three-dimensional hills of moderate slope. Flow also accelerates in going around the hillsides. These features, as well as deceleration in the near wake, are shown in Fig. 14.18, based on measurements over an approximately circular hill (Brent Knoll).

Another feature of the air flow over a hilltop is that the wind speed increases with height much more rapidly than it does over a level ground. Above this shallow surface shear layer, the wind profile becomes nearly uniform; sometimes it is characterized by a maximum wind speed (low-level jet) near the surface. Wind profiles at the tops of escarpments are also found to have similar characteristics.

Flow at an oblique angle to a long ridge is often characterized by the generation of trailing vortices and a persistent swirling turbulent flow in the wake. Thus, the ridge acts like a vortex generator. These and other three-dimensional effects of hills are poorly understood, and so are the Coriolis effects of the earth's rotation on flow around hills.

Very few systematic studies have been made on groups of hills and valleys. The simplest arrangement is that of similar, nearly two-dimensional ridges and valleys located parallel to each other. Wind tunnel experiments indicate that the speed-up of flow over the ridge top and the flow distortion in the near wake are maximum for the first ridge and valley, as if downwind ridges did not exist. But these effects diminish in the downwind direction as subsequent ridges and valleys interact with the flow. After the first few ridges and valleys, the speed-up factor over the tops of ridges approaches a constant value of only slightly above unity, and the summit velocity profile assumes a new logarithmic form with nearly the same roughness length as upwind. Turbulence intensities over the hill crests do not differ much from their undisturbed (flat terrain) values except for some enhancement of the lateral component σ_v/U for wind directions not normal to ridges and valleys. In the valleys between the ridges mean flow is reduced, with possible recirculation cavities forming over steeper lee-side slopes, and turbulence is considerably enhanced. These results have been confirmed in a recent field study of atmospheric flow over a succession of ridges and valleys under near-neutral conditions (Mason and King, 1984). When the flow is across the valley, wind speeds in the valley are about one- to two-tenths of that on the summit. The size of the separated region is very sensitive to flow direction and the presence of any sharp discontinuity in the terrain profile. When the flow is predominantly along the valley, wind speeds in the valley are higher, but still less than that on the summit, and no channeling effect is observed, at least in near-neutral conditions.

A grouping of three-dimensional hills can produce complex flow patterns, similar to those for groups of buildings. For example, channeling of flow between two round hills or in a gap between a long ridge often leads to strong, persistent winds. These effects are particularly pronounced in the presence of stable stratification, which forces the flow to go around rather than over the hills.

14.7.2   STABLY STRATIFIED APPROACH FLOW

Effects of topography on the flow are considerably modified in the presence of stable stratification. Some of the effects and the associated flow patterns are qualitatively similar to those for the neutral flow, although stratification does influence their intensity. Other effects are peculiar to stably stratified flows over and around hills, e.g., the generation of lee waves, formation of rotors and hydraulic jump, inability of the low-level fluid to go over the hills, and upstream blocking.

Stratification effects on flow around topography are generally described in terms of a Froude number, (F or F_L) based on the characteristic height (H) or length scale (L_1) of the topographical feature in the direction of flow

$$F = U_0/NH$$
$$F_L = U_0/NL_1$$

(14.11)

where U_0 is the characteristic velocity of approach flow (say, at $z = H$), and N is the Brunt–Vaisala frequency

$$N \equiv \left(- \frac{g}{\rho_0} \frac{\partial \rho}{\partial z} \right)^{1/2} = \left(\frac{g}{\Theta_0} \frac{\partial \Theta}{\partial z} \right)^{1/2} \qquad (14.12)$$

which is the natural frequency of internal gravity waves or lee waves; the corresponding wavelength is $\lambda = U_0/2\pi N$. Physically, the Froude number is the ratio of inertia to gravity forces determining the flow over topography. It has an inverse-square relationship to the bulk Richardson number, i.e., $F \sim \mathrm{Ri}_B^{-2}$. Note that if $F \ll 1$ the stratification is considered to be strong, while for $F \gg 1$ it is near neutral ($F = \infty$ for a strictly neutral stability). The Froude number also provides the criteria for the possible generation of lee waves and separation of flow in the lee of the hill. In particular, simple Froude number criteria have been developed from the-

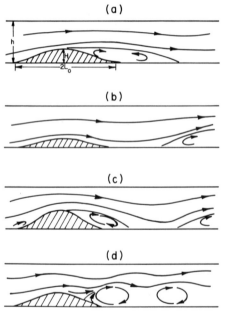

Fig. 14.19 Stratified flows over two-dimensional hills in a channel or under a strong inversion, showing the effect of lee waves on flow separation. (a) Supercritical $F > F_c$, no waves possible; separation is boundary layer controlled. (b) Hill with low slope: subcritical $F < F_c$, downstream separation caused by lee-wave rotor. (c) Hill with moderate slope: supercritical $F > F_c$, boundary layer separation on lee slope. (d) Hill with moderate slope: subcritical $F < F_c$, lee-wave-induced separation on lee slope. [After Hunt and Simpson (1982).]

ory and experiments in the simple case of a uniform, inviscid approach flow with constant density or potential temperature gradient (see, e.g., Hunt and Simpson, 1982). These are schematically shown in Fig. 14.19 for two-dimensional hills. Here, the critical Froude number (F_c) for separation represents the highest value of F at which the boundary layer separation is suppressed by lee waves. It depends on the stratification of the approach flow, as well as on the various hill parameters (e.g., height, aspect ratio, and shape).

Figure 14.19 also indicates that different types of flow patterns may develop in the lee of a two-dimensional hill, depending on the Froude number and the maximum hill slope. Similar and other types of flow phenomena are found to occur in the lee of three-dimensional hills and under different approach flow conditions (e.g., stratified shear flow and mixed layer capped by an elevated inversion). Perhaps the most spectacular flow phenomena related to topography are the severe downslope winds, locally known as Chinook, Föhn, or Bora, followed by a "hydraulic jump." These are analogous to the passage of water over a dam spillway as it is rushing down at very high speeds and creating a violent hydraulic jump at the base of the dam. The necessary approach flow conditions for the corresponding atmospheric flow are a strong, elevated inversion capping a high-speed mixed-layer flow of depth $h > H$, such that $U_0/N(h - H) \simeq 1$, where N is the Brunt–Vaisala frequency for the inversion layer. Figure 14.20 shows the schematics of two widely different, but possible, flow patterns with an approach mixed-layer flow. Note that, here, the relevant Froude number determining the flow over the hill is $U_0/N(h - H)$.

An important aspect of stably stratified flows over and around three-dimensional topographical features is the increasing tendency of fluid parcels to go around rather than over the topography with increased stratification (decreasing Froude number). This is because such fluid parcels do not possess sufficient kinetic energy to overcome the potential energy required for lifting the parcel through a strong, stable density gradient. Whether a given parcel in the approach flow would go over or around the hill depends on the height of the parcel relative to hill height, its initial lateral displacement from the central (stagnation) streamline, shear and stratification in the approach flow, and topographical parameters. Parcels approaching the hill at low heights are likely to go around, while those near the hilltop may go over the hill. One can define a dividing streamline that separates the flow passing around the sides of the hill from that passing over the hill. The height H_s of this dividing streamline can be estimated from the simple criterion that the kinetic energy of a fluid parcel following this streamline be equal to the potential energy associated with

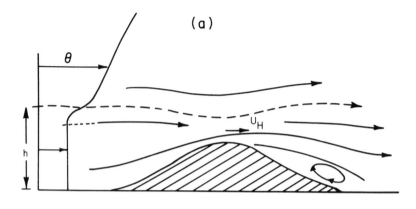

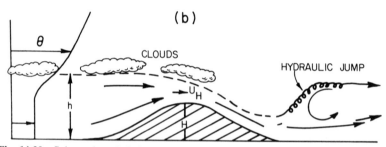

Fig. 14.20   Schematics of air flow over a two-dimensional ridge with an elevated inversion upwind. (a) Low wind speed with $U_0/N(h - H) \ll 1$, flow separation on lee slope. (b) High wind speed with $U_0/N(h - H) \simeq 1$, no flow separation on lee slope, hydraulic jump downwind. [After Hunt and Simpson (1982).]

stratification that the parcel must overcome, i.e.,

$$\frac{1}{2} \rho U_0^2(H_s) = g \int_{H_s}^{H} (H - z)\left(- \frac{\partial \rho}{\partial z}\right) dz \qquad (14.13)$$

This integral equation, first suggested by Sheppard (1956), can be used for any approach flow and topographical arrangement. In practice, it must be solved for H_s iteratively, because the unknown appears as the lower limit of integration. For the particular case of a stratified approach flow with a constant density gradient, Eq. (14.13) reduces to the simpler formula

$$H_s = H(1 - F) \qquad (14.14)$$

which was suggested by Hunt and Snyder (1980) on other theoretical and experimental bases.

Laboratory experiments and a few observations of smoke plumes impinging on hillsides have confirmed the validity and usefulness of the dividing streamline concept (see, e.g., Snyder *et al.*, 1985). These also indicate that Eq. (14.13) provides only a necessary (but not the sufficient) condition for the fluid above H_s to go over rather than around a hill. It can still pass around the sides and take a path requiring less potential energy to overcome than that given by Eq. (14.13). Experiments also indicate that the lateral aspect ratio (W/H) of the hill is relatively unimportant, but the stratification and shear in the approach flow, upwind slope of the hill, and obliqueness of the flow to an elongated hill are all important in determining the flow over and around the hill.

In the case of strongly stratified ($F \ll 1$) flow approaching normal to a very long (nearly two-dimensional) ridge, there is a distinct possibility of upstream blocking of the fluid, because very little can go over the ridge. A region of nearly stagnant fluid occurs below the dividing streamline and far away from the edges where it can go around the ridge. Such a flow field may not acquire a steady state, because the upwind edge of the blocked flow propagates farther upstream as a density front and certain columnar disturbance modes and gravity waves generated near the ridge can also propagate upstream. Topographically blocked flows have important implications for local weather and for dispersion of pollutants from upwind sources.

14.8 APPLICATIONS

Spatial variations of surface roughness, temperature, wetness, and elevation frequently cause nonhomogeneous atmospheric boundary layers developing over these surface inhomogeneities. A basic understanding of horizontal and vertical variations of mean flow, thermodynamic variables, and turbulent exchanges in nonhomogeneous boundary layers is essential in micrometeorology and its applications in other disciplines and human activities. Specifically, the following applications may be listed for the material covered in this chapter:

- Determining the development of modified internal boundary layers following step changes in surface roughness and temperature
- Determining the fetch required for air flow adjustment to the new surface following a discontinuity in surface elevation or other properties
- Selecting an appropriate site and height for locating meteorological or micrometeorological instruments for representative observations for the local terrain

- Determining wind and air mass modifications under situations of cold or warm air advection over a warmer or colder surface
- Estimating urban influences on surface energy balance and surface and air temperatures
- Determining possible modifications of mean flow, turbulence, and the PBL height over an urban area
- Estimating transport and diffusion of pollutants in the urban boundary layer and the urban plume
- Estimating pollutant dispersion around hills and buildings
- Estimating wind loads on buildings and knowing the wind environment around buildings
- Siting and designing wind-power generators in hilly terrain
- Other wind-engineering applications in urban architecture and planning

PROBLEMS AND EXERCISES

1. A 1500-m thick near-neutral atmospheric boundary layer encounters a sudden change in the surface roughness in going from land ($z_0 = 0.1$ m) to a large lake ($z_0 = 0.0001$ m) of the same surface temperature. The observed wind speed at a 10-m height in the approach flow is 10 m sec^{-1}. Assuming that the overall PBL thickness ($h = 1500$ m) and the surface layer thickness ($h_s \simeq 150$ m) remain unchanged over the lake, calculate and plot the following parameters, using Eqs. (14.1) and (14.3), with $a_i = 0.4$ for the internal boundary layer:
   (a) The IBL height as a function of distance, and the distances where it reaches the top of the approach flow surface layer and the PBL
   (b) The friction velocity and wind speed at a 10-m height over the lake as functions of distance from the shore
2. Repeat the calculations in Problem 1 for the case of an onshore neutral boundary layer flow with the same wind speed (10 m sec^{-1}) and depth (1500 m) over the lake, and compare the results for the two cases.
3. Calculate and graphically compare the development of thermal internal boundary layers in the following two situations, using Eqs. (14.8c) and (14.8d), respectively, with $a_4 = 1.7$ and $a_5 = 1.0$.
   (a) A stably stratified boundary layer flow with $\partial\Theta/\partial z = 0.02$ K m^{-1} over land advecting over a 10°C warmer sea. Assume a typical value of the drag coefficient $C_D \equiv (u_*/U_m)^2 \simeq 0.002$ for the free convective marine boundary layer.
   (b) An unstable warm continental boundary layer capped by an inver-

sion with $\partial T/\partial z = 0.08$ K m^{-1} advecting over a 10°C colder sea surface. Assume a typical drag coefficient of 0.5×10^{-3} for the stably stratified IBL.

4. (a) What are the leading causes of an urban heat island?
   (b) If the population of a midwestern United States city has increased from 100,000 to 1,000,000 in the past 20 yr, estimate the expected change in its maximum heat-island intensity and the annual rate of urban warming over that period.

5. (a) Give physical reasons for the urban mixed-layer dome phenomenon and list conditions under which it might be enhanced and those in which it would be suppressed.
   (b) Calculate and compare the nocturnal urban mixed-layer height at the city center, where the near-surface temperature is 15°C, with that at a suburban location of near-surface temperature 12°C, when the upwind rural sounding indicates a temperature gradient of 0.02°C m^{-1} over a deep-surface inversion layer and a near-surface temperature of 9°C.

6. (a) What are the causes and consequences of boundary layer separation from hills and buildings?
   (b) In what respects do the cavity and near-wake flow regions behind an isolated square-shaped building differ from those behind a cylindrical building?

7. (a) Why is the length of a recirculating cavity region behind a long ridge much larger than that behind an axisymmetrical hill of the same height and slope?
   (b) In what respects does a three-dimensional hill wake differ from a building wake?

8. (a) Show that, for constant density-gradient uniform or shear flow, Eq. (14.13) reduces to Eq. (14.14).
   (b) Calculate the dividing streamline height for a constant gradient shear flow with $\partial U_0/\partial z = 0.1$ sec^{-1}, $\partial \Theta/\partial z = 0.05$ K m^{-1}, and $\Theta_0 = 280$ K, approaching a 200-m-high axisymmetric hill. What are the implications of this to flow over and around the hill?

Chapter 15 | Agricultural and Forest Micrometeorology

15.1 FLUX-PROFILE RELATIONS ABOVE PLANT CANOPIES

In this final chapter we discuss the particular applications of micrometeorological principles and methods to vegetative surfaces, such as agricultural crops, grasslands, and forests. Here, the micrometeorologist is interested in the atmospheric environment, both above and within the plant canopies, as well as in the complex interactions between vegetation and its environment. First, we discuss the vertical profiles of mean wind, temperature, specific humidity, etc., and their relationships to the vertical fluxes of momentum, heat, water vapor, etc., above plant canopies.

The semiempirical flux-profile relations developed in Chapters 10–12 should, in principle, apply to any homogeneous surface layer, well above the tops of roughness elements ($z > h_0 \gg z_0$). It should be kept in mind, however, that for any tall vegetation the apparent reference level for measuring heights must be shifted above the true ground level by an amount equal to the zero-plane displacement (d_0), which is related to the mean height (h_0) of roughness elements and roughness density. Also, for flexible plant canopies, z_0 and d_0 may not have fixed ratios to h_0 but may vary with wind speed. Physically, one can argue that for a vegetative canopy acting as a sink of momentum and a source or sink of heat, water vapor, CO_2, etc., the effective source or sink height must lie between 0 and h_0. For exchange processes in the atmospheric boundary layer well above the canopy, it is found convenient to consider the canopy as an area source or sink of infinitesimal thickness located at an effective height d_0 above the ground level, and to consider it an effective reference plane. Then, the zero-plane displacement does not appear explicitly in theoretical or semiempirical flux-profile relations, such as those given in Chapters 10–12. In most experimental situations and practical applications, however, the ground level is the most convenient reference plane for measuring heights. It is also the appropriate reference plane for representing mean and turbulent quantities within the canopy layer. For these reasons, in this chapter, we will use the height z above the local ground level and account for the zero-plane displacement, wherever appropriate (e.g., in above-canopy flux-profile relations).

Applying the Monin–Obukhov similarity relations to the fully developed (horizontally homogeneous) above-canopy surface layer, the mean wind speed (U), the potential temperature (Θ), and the specific humidity (Q), profiles can be represented by

$$\frac{U}{u_*} = \frac{1}{k}\left[\ln\left(\frac{z - d_0}{z_0}\right) - \psi_m\left(\frac{z - d_0}{L}\right)\right]$$

$$\frac{\Theta - \Theta_0}{\theta_*} = \frac{1}{k}\left[\ln\left(\frac{z - d_0}{z_0}\right) - \psi_h\left(\frac{z - d_0}{L}\right)\right] \qquad (15.1)$$

$$\frac{Q - Q_0}{q_*} = \frac{1}{k}\left[\ln\left(\frac{z - d_0}{z_0}\right) - \psi_w\left(\frac{z - d_0}{L}\right)\right]$$

where Θ_0 and Q_0 are the values at $z = d_0 + z_0$, and θ_* and q_* are the scaling parameters, each representing the ratio of the appropriate vertical flux (in kinematic units) to the friction velocity u_* [see, e.g., Eq. (12.12)]. The M–O similarity functions ψ_h and ψ_w are presumably equal, but may differ from the momentum function ψ_m, especially in unstable and free convective conditions. The empirical forms of these functions have already been discussed in Chapters 11 and 12.

The gradient and bulk parameterizations of the vertical fluxes of momentum, heat, etc., in an idealized (constant-flux, horizontally homogeneous, quasistationary) surface layer have also been discussed in Chapters 11 and 12. These can be applied to the above-canopy surface layer, especially in the height range $0.1h > z > 1.5h_0$. The range of validity of the surface layer similarity relations becomes narrower for taller canopies and for shallow boundary layers. It may disappear completely when the PBL thickness is less than 10–15 times the vegetation height, a distinct possibility over mature forest canopies.

TRANSFER COEFFICIENTS AND RESISTANCES

The flux parameterizations that employ the drag coefficient and other transfer coefficients are commonly used in micrometeorology and its applications to air–sea interaction, air pollution, hydrology, large-scale atmospheric modeling, etc. In these, the various fluxes are linearly related to the products of wind speed and the difference between the values of variables at some reference height in the surface layer and at the surface itself. For example

$$\tau_0 = \rho C_D U^2$$

$$H_0 = \rho c_p C_H U(\Theta_0 - \Theta) \qquad (15.2)$$

$$E_0 = \rho C_W U(Q_0 - Q)$$

in which the drag coefficient (C_D) and other transfer coefficients can be prescribed as functions of the surface roughness (more appropriately, z/z_0) and stability parameters (z/L or Ri_B), as discussed in Chapter 11. On the basis of Reynolds analogy (similarity of transfer mechanisms of heat, water vapor, etc.), one expects that $C_H = C_W$.

Another commonly used approach or method of parameterizing the vertical fluxes in terms of the difference in mean property values at two heights is the "aerodynamic resistance" approach, which is based on the analogy to electrical resistance and Ohm's law:

$$\text{current} = \frac{\text{potential difference}}{\text{electrical resistance}}$$

The analogous relations for turbulent fluxes are

$$\tau_0 = \rho r_M^{-1} U$$

$$H_0 = \rho c_p r_H^{-1}(\Theta_0 - \Theta) \qquad (15.3)$$

$$E_0 = \rho r_W^{-1}(Q_0 - Q)$$

in which r_M, r_H, etc., are the aerodynamic or aerial resistances to the transfer of momentum, heat, etc. Note that a flux will increase as the appropriate aerodynamic resistance decreases.

The resistance approach is more frequently used in micrometeorological applications in agriculture and forestry. This is largely because of the simple additive property of resistances in series, which can be used to express the total aerodynamic resistance in terms of resistances of its components (bare soil surface, leaves, branches, and stems of plants). A disadvantage is that, unlike the dimensionless transfer coefficients, aerodynamic resistances are dimensional properties, having dimensions of the inverse of velocity, which depend not only on the canopy structure, but also on the wind speed at the appropriate height and atmospheric stability. Also, in practice, it is not easy to determine and specify the component resistances for estimating the overall canopy resistance. A comparison of Eqs. (15.2) and (15.3) clearly shows that the bulk transfer method is clearly superior to the aerodynamic resistance method, but the two sets of coefficients are interrelated as

$$r_M^{-1} = UC_D; \qquad r_H^{-1} = UC_H; \qquad r_W^{-1} = UC_W \qquad (15.4)$$

Thus, for a given canopy, aerial resistances are inversely related to wind speed and are not unique properties of canopy and the surface.

15.2 RADIATION BALANCE WITHIN PLANT CANOPIES

Solar or shortwave radiation is the main source of energy for the vegetation. Longwave radiation is also a major component of energy balance in daytime and the primary component at nighttime. In plant canopies radiation has important thermal and photosynthetic effects; it also plays a major role in plant growth and development processes. The visible part of the shortwave radiation (0.38–0.71 μm) is called the photosynthetically active radiation (PAR).

Radiative transfer in a plant canopy is a very complicated problem for which no satisfactory general solution has been found. There are significant amounts of internal radiative absorption, reflection, transmission, and emission within the canopy elements. Complications also arise from the large variability and inhomogeneity of the canopy architecture, as well as of individual plants. The canopy architecture also determines, to a large extent, the turbulent exchanges of momentum, heat, water vapor, CO_2, etc., within the canopy and is discussed briefly in the following section.

15.2.1 CANOPY ARCHITECTURE

The size and shape of canopy elements, as well as their distributions in space and time, determine the physical characteristics of the canopy structure. A detailed description of a nonhomogeneous or nonuniform plant canopy in a complex terrain would necessarily involve many parameters. A commonly accepted simplification is made by assuming that the plant stand is horizontally uniform and its average characteristics may vary only with height above the ground.

The simplest characteristic of the canopy structure is its average height or thickness h_0. Along with the area density of the plants, h_0 determines the roughness parameter and the effective heights of the sources or sinks of momentum, heat, moisture, etc. Another widely used characteristic of the canopy architecture is the cumulative leaf area index (L_{AI}), defined as the area of the upper sides of leaves within a vertical cylinder of unit cross-section and height h_0. The leaf area index is a dimensionless parameter which is related to the foliage area density function $A(z)$ as (Ross, 1975)

$$L_{AI} = \int_0^{h_0} A(z) \, dz \qquad (15.5)$$

Here, $A(z)$ is a local characteristic of the canopy architecture which represents the one-sided leaf area per unit volume of the canopy at the height z. A more convenient dimensionless local characteristic is the local cumulative leaf area index

$$L_{AI}(z) = \int_z^{h_0} A(z) \, dz \qquad (15.6)$$

which represents the leaf area per unit horizontal area of the canopy above the height z. Note that $L_{AI}(0) = L_{AI}$.

In most crop and grass canopies leaves are the most active and dominant elements of interaction with the atmosphere, and the fractional area represented by plant stems and branches can be neglected. Exceptions are certain forest canopies in which woody elements may not be ignored. Particularly for a deciduous forest in late fall and winter, the woody-element silhouette area index (W_{AI}), which can be defined in a manner similar to L_{AI}, is an important architectural characteristic. The sum of the two indices is called the plant area index

$$P_{AI} = L_{AI} + W_{AI} \qquad (15.7)$$

Both the leaf area density and the leaf area index are found to vary over a wide range for the various crops and forests. For a particular crop, they also depend on the density of plants and their stage of growth. For illustration purposes, Figs. 15.1 and 15.2 show their observed distributions in a deciduous forest canopy in summer months. Note that for this particular canopy, $L_{AI} \simeq 4.9$, $W_{AI} \simeq 0.6$, and $P_{AI} \simeq 5.5$. Observations of the same parameters for other forest and crop canopies are reviewed in Monteith (1975, 1976).

The orientation of leaves is characterized by the mean leaf inclination angle from the horizontal. One can also measure the cumulative frequency distribution of the inclination angle. The foliage of many plant stands are more or less uniformly oriented, but some may display a narrow range of inclination angles around the horizontal, vertical, or somewhere in between (these are called panophile, erectophile, and plagiophile canopies). A useful inclination index of the foliage is defined by Ross (1975). This index also depends on plant species and has a height and seasonal dependency.

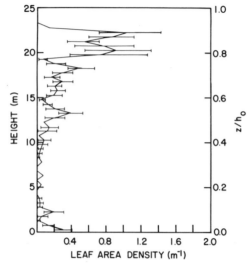

Fig. 15.1 Vertical distribution of mean leaf area density in a deciduous forest in eastern Tennessee. [After Hutchison *et al.* (1986).]

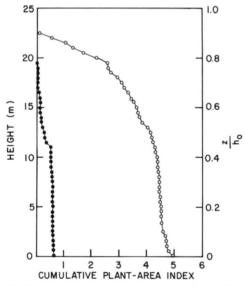

Fig. 15.2 Vertical distribution of local cumulative leaf area index (○) and woody element area index (●) in a deciduous forest in eastern Tennessee. [After Hutchison *et al.* (1986).]

15.2.2   PENETRATION OF SHORTWAVE RADIATION

For a given incident solar radiation at the top of the canopy, the radiation regime in a plant canopy is determined by the canopy architecture, as well as by radiative properties of the various canopy elements and the ground surface. At any particular level in the canopy, the incoming solar radiation (both direct and diffuse) may vary considerably in the horizontal due to the presence of sunflecks, and shadow areas with transitional areas, called penumbra, in between. Therefore, appropriate spatial averaging is necessary when we represent the radiative flux density as a function of height alone.

The transmission of incoming shortwave radiation into a plant canopy shows an approximately exponential decay with depth of penetration (more appropriately, the depth-dependent leaf area index), following the Beer–Bouguer law

$$R_{S\downarrow}(z) = R_{S\downarrow}(h_0) \exp[-\alpha L_{AI}(z)] \tag{15.8}$$

in which α is an extinction coefficient. The same law, perhaps with a slightly different value of extinction coefficient, applies to the net shortwave radiation. A comparison of observed profiles of incoming and net shortwave radiations in a grass canopy with Eq. (15.8) is shown in Fig. 15.3. The diurnal variation of the extinction coefficient for the same canopy is shown in Fig. 15.4. Radiation measurements in forest canopies also show the approximate validity of Eq. (15.8), with P_{AI} replacing L_{AI}.

Foliage in a canopy not only attenuates the radiative flux density but it also changes its spectral composition. Leaves absorb visible (PAR) radiation more strongly than they absorb the near infrared radiation (NIR). This selective absorption reduces the photosynthetic value of the solar radiation as it penetrates through greater depths in the canopy. The spectral composition of total radiation also differs in sunflecks and shaded areas; in sunflecks the spectrum is similar to that of the incident total radiation, while in shaded areas the NIR dominates.

The shortwave reflectivity or albedo of a plant canopy depends on the average albedo of individual leaves, the canopy architecture (in particular, the canopy height, the leaf area index, and the leaf inclination index), and the solar altitude angle. The latter two factors determine the amount of penetration, radiation trapping, and mutual shading within the canopy. The longer the path of attenuation, the greater the amount of radiation to be trapped and the less to be reflected. Consequently, the albedo of most plant canopies is a decreasing function of both the canopy height and the solar altitude (see, e.g., Oke, 1978, Chapter 4).

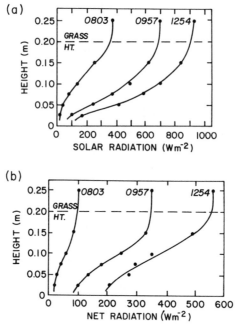

Fig. 15.3 Observed profiles of (a) incoming solar radiation ($R_{S\downarrow}$) and (b) net all-wave radiation (R_N) in a 0.2-m stand of native grass at Matador, Saskatchewan, on a clear summer day. [From Oke (1987); after Ripley and Redmann (1976).]

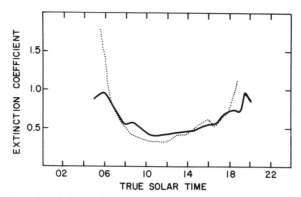

Fig. 15.4 Diurnal variations of incoming solar and net radiation extinction coefficients for the native grass at Matador, Saskatchewan. ·····, Net radiation; ——, total shortwave radiation. [After Ripley and Redmann (1976).]

15.2.3 LONGWAVE RADIATION

The total or net longwave radiation (R_L) at any height within a plant canopy has four components: (1) the longwave radiation from the atmosphere which has penetrated through the upper layer of the canopy without interception; (2) the longwave radiation from the upper leaves and other canopy elements; (3) the radiation from lower canopy elements; and (4) the part of outgoing longwave radiation from the ground surface which is not intercepted by lower canopy elements. The net longwave radiation at the top of a plant canopy is almost always a loss, as with most other surfaces. However, the net loss usually diminishes toward the ground due to the reduction in the sky view factor (SVF), as the cold sky radiative sink is increasingly being replaced by relatively warmer vegetation surfaces. Consequently, in the lower part of a tall and dense canopy the role of sky becomes insignificant and R_L depends largely on the ground-surface temperature, the vertical distribution of mean foliage temperature, and ground and leaf surface emissivities. If the foliage temperature distribution is nearly uniform, which may be the usual case, R_L is also approximately constant, independent of height. The variation of R_L in the upper parts of the canopy is much more complicated (see, e.g., Ross, 1975).

15.3 WIND DISTRIBUTION IN PLANT CANOPIES

Wind profiles above plant canopies ($z \geq 1.5h_0$) have been observed to follow the log law or the modified log law, depending on the atmospheric stability. Just above and within the canopy, however, the observed profiles deviate systematically (toward higher wind speeds) from the theoretical (modified log law) profile. The actual flow around roughness elements being three-dimensional, we can think of some generalized vertical profiles of wind and momentum flux only in a spatially averaged sense. With mechanical mixing processes dominating within the canopy, the stability effects can safely be ignored. If the large-scale pressure gradients can also be ignored, as is commonly done in the above-canopy surface layer, the equations of motion degenerate to the usual condition that the divergence of the total vertical momentum flux must be zero. The main difference between the flow inside the canopy and that above it is that considerable momentum is absorbed over the depth of the canopy in the form of drag due to the various canopy elements.

An appropriate equation for horizontally averaged wind flow within canopy is (Plate, 1971; Businger, 1975)

$$\partial \overline{uw}/\partial z = -(1/\rho)(\partial P/\partial x) - \tfrac{1}{2}C_d A U^2 \tag{15.9}$$

where C_d is the average drag coefficient of the plant elements and A is the effective aerodynamic surface area of the vegetation per unit volume. Both C_d and A are likely to be complicated functions of height and the canopy structure. When the pressure term is negligible, e.g., in the upper part of the canopy, Eq. (15.9) reduces to

$$\partial \overline{uw}/\partial z = -\tfrac{1}{2}C_d A U^2 \qquad (15.10)$$

Practical models of wind flow in canopy are based on the solution to Eq. (15.10) with appropriate assumptions for C_d, A, and the relation between shear stress and mean velocity gradient (e.g., an eddy viscosity or a mixing-length hypothesis). Using the simplest assumption that $C_d A$ is independent of height and the mixing-length relation [Eq. (9.16)] is valid with a constant l_m, the solution to Eq. (15.10) is an exponential wind profile (Inoue, 1963)

$$U(z)/U(h_0) = \exp[-n(1 - z/h_0)] \qquad (15.11)$$

where $U(h_0)$ is the wind speed at the top of the canopy and $n \equiv (C_d A/4 l_m^2)^{1/3} h_0$.

An exponential profile similar to Eq. (15.11) can also be derived from an eddy viscosity relation for momentum flux with an exponential variation of K_m which is consistent with a constant l_m (see, e.g., Thom, 1975). If, on the contrary, K_m is assumed constant, the resulting canopy wind profile is of the form

$$U(z)/U(h_0) = [1 + m(1 - z/h)]^{-2} \qquad (15.12)$$

where m is a numerical coefficient characterizing the plant canopy. The actual difference in the profile shapes given by Eqs. (15.11) and (15.12) is small for the appropriate choice of profile parameters n and m (compare the two, for example, for $n = m = 3$), despite the strong contrast in the associated K_m profiles. This seems to indicate that the predicted canopy wind profile is not very sensitive to the assumed K_m or l_m distribution within reasonable limits.

The simplifications and assumptions used in the derivation of the exponential profile [Eq. (15.11)] are such that one may wonder if it could represent the observed wind profiles in plant canopies. Surprisingly, however, a number of observed profiles do agree with Eq. (15.11), especially in the upper half of canopies. This is partially because the parameter n is only weakly dependent on C_d and A. Figure 15.5 shows a comparison of some observed profiles in wheat and corn canopies with the exponential profile Eq. (15.11). The data suggest that a unique (similarity) profile shape may exist for each canopy type. Cionco (1972) has summarized the best estimated empirical values of n for various canopies; they usually lie

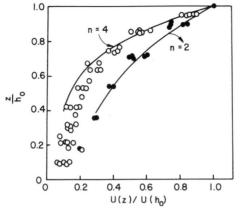

Fig. 15.5 Comparison of observed mean velocity profiles in wheat (○) and corn (●) canopies with the exponential profile [Eq. (15.11), with $\alpha = 4$ and 2, respectively. [After Plate (1971); Businger (1975).]

between 2 and 4. This simple theoretical or semiempirical model may represent only a part of the canopy wind profile. The model can be refined further by considering variations with height of C_d, A, and l_m (see, e.g., Cionco, 1965).

In a tall and well-developed canopy such as a mature forest, most of the momentum is absorbed in the upper part of the canopy where leaves are concentrated. Farther down, as the leaf area density decreases to zero, the pressure gradient term in Eq. (15.9) is likely to become more significant and perhaps dominate over the flux-divergence term. Consequently, wind speed may increase with decreasing height above the ground, until surface friction reverses this trend close to the ground level. This suggests that a low-level wind maximum (jet) may appear in the lower part of certain forest canopies. This is indeed confirmed by many observations, a sample of which are shown in Fig. 15.6. Note that a variety of wind profile shapes are found in forest canopies.

15.4 TEMPERATURE AND MOISTURE FIELDS

Temperature profiles have been measured in various crop and forest canopies; these are quite different from those above the canopies. There are differences in canopy temperature profiles due to different canopy architectures. There are also significant diurnal variations in these due to variations of dominant energy fluxes. Here, we shall discuss only some typical observations and point out some of the common features of can-

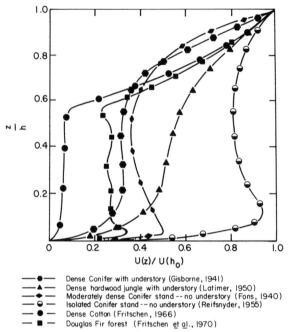

$$\frac{z}{h}$$

$$U(z)/U(h_o)$$

— ● — Dense Conifer with understory (Gisborne, 1941)
— ▲ — Dense hardwood jungle with understory (Latimer, 1950)
— ◆ — Moderately dense Conifer stand -- no understory (Fons, 1940)
— ◖ — Isolated Conifer stand -- no understory (Reifsnyder, 1955)
— ● — Dense Cotton (Fritschen, 1966)
— ■ — Douglas Fir forest (Fritschen et al., 1970)

Fig. 15.6 Observed mean wind profiles in different forest canopies. [From Businger (1975); after Fritschen *et al.* (1970).]

opy temperature and humidity profiles. For a more comprehensive review of the literature the reader should refer to Monteith (1975, 1976).

Figure 15.7 shows the observed hourly averaged air temperature profiles within and above an irrigated soybean crop over the course of a day. Often, during daytime, there is a temperature maximum near middle to upper levels of the canopy. Since the sensible heat exchange in the canopy is expected to be down the temperature gradient, the heat flux is downward (negative) below the level of temperature maximum and upward (positive) above it. Thus, the sensible heat diverges away from the source region which coincides with the level of maximum leaf area density where most of the solar radiation is absorbed. The temperature inversion in the lower part of the canopy is a typical feature of daytime temperature profiles in tall crop and forest canopies. At night, temperature profiles in lower parts of canopies are close to isothermal, because canopies trap most of the outgoing longwave radiation. Often, the minimum temperature occurs not at the ground level but just below the crown. The nocturnal temperature inversion extends from above this level. The tem-

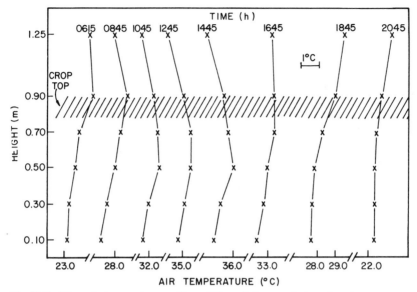

Fig. 15.7 Mean air temperature profiles within and above an irrigated soybean crop on a summer day at Mead, Nebraska. [After Rosenberg *et al.* (1983).]

perature gradient or inversion strength above a canopy is usually much smaller than that over a bare ground surface.

Figure 15.8 gives the hourly mean profiles of wind speed, temperature, water vapor pressure, and carbon dioxide inside and above a barley canopy on a summer day for which the observed canopy energy budget is given in Fig. 2.4. The observed wind and temperature profiles display the same features as discussed above. The daytime profiles of water vapor pressure show the expected decrease with height through much of the canopy and the surface layer above; a slight increase (inversion) in the upper part of the canopy is probably due to increased transpiration from the leaves in that region. Note that both the soil surface and foliage are moisture sources. The nocturnal vapor pressure profiles are more complicated by the processes of dewfall and guttation on leaves, while weak evaporation may continue from the soil surface.

Temperature and water vapor pressure profiles in forest canopies are similar to those within tall crops. The main difference is that the gradients are much weaker and diurnal variations are smaller in forest canopies. Even on warm, sunny days, air is cooler and more humid inside forests. Figure 15.9 presents the typical daytime profiles of meteorological variables and foliage area density. Both the temperature and the water vapor

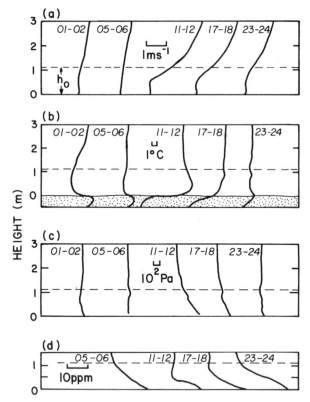

Fig. 15.8 Observed profiles of (a) wind speed, (b) temperature, (c) water vapor pressure, and (d) carbon dioxide concentration in and above a barley crop on a summer day at Rothamsted, England. [From Oke (1987); after Long *et al.* (1964).]

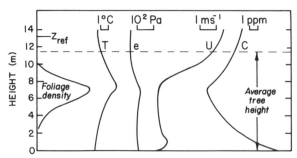

Fig. 15.9 Observed mean profiles of foliage density and meteorological variables in a Sitka spruce forest on a sunny summer day near Aberdeen, Scotland. [After Jarvis *et al.* (1976).]

pressure profiles show maximum values at the level of maximum foliage density where the sources of heat (radiative absorption) and water vapor (transpiration) are most concentrated. Beneath this there is a temperature inversion, but nearly uniform vapor pressure profile; an inversion in the latter would be expected if the forest floor were dry. At night the temperature profile (not shown) is reversed, with a minimum occurring near the level of maximum foliage density. Due to reduced sky view factor and trapping of longwave radiation in the lower part of the canopy, temperatures remain mild there, compared to those in open areas. Dewfall, if any, is largely confined to the upper part of the canopy, just below the crown. In the absence of dewfall, specific humidity or water vapor pressure is a decreasing function of height with weak evaporation from the soil surface as the only source of water vapor in the canopy.

15.5   TURBULENCE IN PLANT CANOPIES

The generation of turbulence in a plant canopy is a much more complex process than that over a flat bare surface or in a homogeneous boundary layer above a plant canopy. In addition to the usual sources of turbulence generation, such as mean wind shear and buoyancy, each canopy element is a source of turbulence due to its interaction with the flow and generation of separated shear layer and wake. The wake-generated turbulence in the canopy has characteristic large-eddy scales comparable to the characteristic sizes of the various canopy elements (e.g., leaves, branches and trunk). These scales are usually much smaller than the large-eddy scales of turbulence ($z - d_0$ and $h - d_0$) above the canopy. However, large-scale turbulence also finds its way through the canopy intermittently during short periods of downsweeps bringing in higher momentum fluid from the overlying boundary layer. It has been observed that the bulk of turbulent exchange between the atmosphere and a canopy may be occurring during these large-scale downsweeps and compensating updrafts. Thus, canopy turbulence is distinguished by multiplicity of scales and generating mechanisms. It is also characterized by high-turbulence intensities, with turbulent velocity fluctuations of the same order or even larger than the mean velocity. These characteristics make canopy turbulence not only extremely difficult to measure, but also preclude a simple representation of it in a similarity form. Still, for convenience, the height above the surface is usually normalized by the canopy height and σ_u, σ_v, σ_w are normalized by U or u_*.

15.5.1 TURBULENCE INTENSITIES AND VARIANCES

Limited observations of longitudinal turbulence intensity $i_u = \sigma_u/U$ suggest that, typically, $i_u \simeq 0.4$ in agronomic crops, 0.6 in temperate forests and between 0.7 and 1.2 in tropical forests (Raupach and Thom, 1981). In a gross sense, i_u increases with canopy density. The turbulence intensity profiles broadly follow the leaf or plant area density profiles, but are also influenced by mean wind speed and stability within the canopy. The lateral and vertical turbulence intensities (i_v and i_w) are generally proportional to but smaller than i_u. A comparison of turbulence intensity profiles within rice and maize canopies shows that the profile is nearly uniform in the former, with an approximately uniform leaf area density distribution, while the profile in the maize canopy has a maximum near the top of the canopy (see Uchijima, 1976). More recent measurements of turbulence in a maize canopy also indicate increasing turbulence activity with height. For example, Figure 15.10 shows the observed profiles of normalized standard deviations of horizontal and vertical velocity components.

15.5.2 TURBULENT FLUXES

Direct measurements of turbulent fluxes of momentum, heat, and water vapor are extremely difficult to make in plant canopies, and a few at-

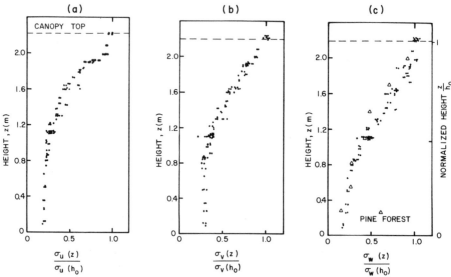

Fig. 15.10 Observed profiles of the normalized standard deviations of (a) longitudinal, (b) lateral, and (c) vertical velocity fluctuations in a corn canopy. [After Wilson *et al.* Copyright © (1982) by D. Reidel Publishing Company. Reprinted by permission.]

tempts have been made only recently. Figures 15.11 and 15.12 show some eddy-correlation measurements of momentum and heat fluxes in a maize canopy. In general, fluxes increase with height and reach their maximum values somewhere near the top of the canopy. Their normalized profiles also show some day-to-day variations.

More often, turbulent fluxes are estimated from the observed mean profiles using gradient-transport hypotheses (e.g., eddy viscosity or mixing length). This approach suffers from many conceptual and practical limitations, however, and its validity has been seriously questioned by micrometeorologists in recent years (see, e.g., Raupach and Thom, 1981). It has been shown that vertical fluxes in plant canopies are not always down the mean vertical gradients, as implied in simpler gradient-transport hypotheses. For example, the vertical momentum flux $\overline{u'w'}$ may be driven by the horizontal gradient of mean vertical velocity ($\partial W/\partial x$) and/or by buoyancy, and need not be uniquely related to $\partial U/\partial z$. Even when it can be related to the latter through an eddy viscosity or a mixing-length relation-

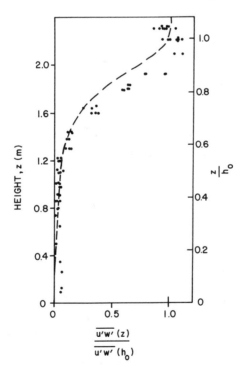

Fig. 15.11 Measured and calculated profiles of normalized shear stress in a corn canopy. [After Wilson *et al.* Copyright © (1982) by D. Reidel Publishing Company. Reprinted by permission.]

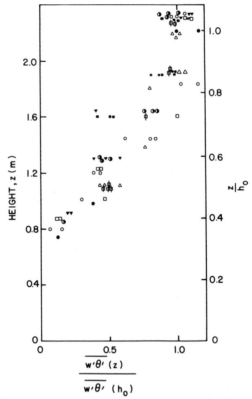

Fig. 15.12 Daily normalized sensible heat flux profiles from eddy correlation measurements in a corn canopy. Day: ●, 1; △, 2; □, 3; ○, 4; ⬚, 5; ■, 6; ▼, 7; ◑, 8. [After Wilson *et al.* Copyright © (1982) by D. Reidel Publishing Company. Reprinted by permission.]

ship, the prescription of K_m or l_m is not straightforward. A number of mutually contradictory K_m and l_m profiles have been proposed in the literature and there is much confusion regarding the shapes of flux profiles. There are very few direct observations of fluxes and gradients in canopy flows which support the gradient-transport (K) theories. On the contrary, observations show strong evidence for countergradient fluxes. Local-diffusion theories cannot cope with and predict such countergradient fluxes. The main reasons for their failure in canopy flow are three-dimensionality of the mean motion, nonlocal production of turbulence, and the multiplicity of the scales of turbulence.

A far better method of estimating turbulent transports or fluxes from measurements of mean wind, temperature, etc., inside canopies is based on the integration of mean momentum and other conservation equations with respect to height. For example, an integration of Eq. (15.10) yields

$$\frac{\overline{uw}(z) - \overline{uw}(0)}{\overline{uw}(h_0) - \overline{uw}(0)} = \frac{\int_0^z C_d A U^2 \, dz}{\int_0^{h_0} C_d A U^2 \, dz} \qquad (15.13)$$

in which the integrals can be numerically evaluated for the given (measured) U profile, the leaf area density $A(z)$, and the average drag coefficient $C_d(z)$. The flux $\overline{uw}(0)$ at the canopy floor can usually be neglected. With this assumption, Eq. (15.13) is compared with the observed profile in Fig. 15.11. The most uncertain parameter in this indirect estimation of momentum flux is the canopy drag coefficient C_d, which is known only for certain idealized plant shapes such as a circular cylinder.

Similar relationships can be derived for sensible heat and water vapor fluxes, using their appropriate conservation equations for the canopy layer. Figures 15.13 and 15.14 show the observed profiles of soil and air temperatures and water vapor pressure, together with the calculated profiles of sensible and latent heat fluxes in a grass canopy ($h_0 \simeq 0.17$ m). Radiative fluxes within this canopy for the same day were given in Fig. 15.3. Note that all the fluxes increase with height and attain their maximum values just above the canopy. During the late morning period the sensible heat flux is negative in the shallow surface inversion layer that gets destroyed later in the afternoon. This inversion in the lower part of the canopy probably produces a build-up of water vapor in this region during the morning period (see Fig. 15.14a). The decrease in vapor pressure in early afternoon is a result of more vigorous turbulent mixing, bringing in drier air from higher levels. The highest vapor pressure occurs at the moist ground surface. The computed latent heat flux profiles in Fig. 15.14b indicate that the major source region is around $z = d_0 \simeq 0.10$ m in the morning and somewhat higher during the afternoon (Ripley and Redman, 1976).

The calculated profiles of eddy diffusivities of heat and momentum based on the above observations in a grass canopy during the morning period are shown in Fig. 15.15. Within the canopy the estimated values of K_h and K_m are in close agreement, indicating the dominance of mechanically generated turbulence in the canopy layer. Above the canopy ($z > h_0 \simeq 0.17$ m), however, K_h values increase more rapidly with height than K_m values, which might be expected from increasing buoyancy effects on turbulence. These are probably the most consistent estimates of eddy diffusivities in a canopy flow, suggesting the usefulness and validity of gradient-transport relations, at least in a thin grasslike canopy. The observation that other periods of the day did not show such a good agreement cautions against any broad generalization of such results.

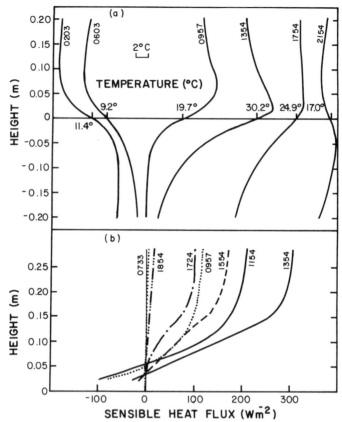

Fig. 15.13   Observed profiles of (a) temperature in the canopy and upper layer of soil and (b) calculated profiles of sensible heat flux in a grass canopy at various times during a summer day. [After Ripley and Redmann (1976).]

15.6   APPLICATIONS

A primary objective of the various micrometeorological studies involving plant canopies has been to better understand the processes of momentum, heat, and mass exchanges between the atmosphere and the biologically active canopy. These exchanges influence the local weather as well as the microclimate in which plants grow. A basic understanding of exchange mechanisms and subsequent development of practical methods of parameterizing the same in terms of more easily and routinely measured parameters is essential for many applications in meteorology, agriculture,

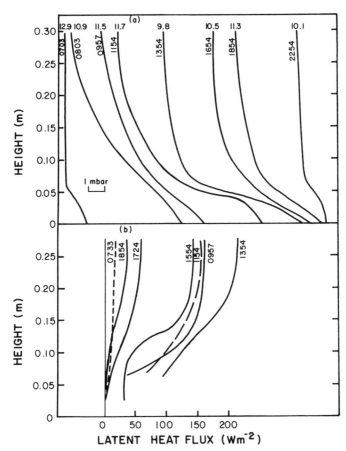

Fig. 15.14 Observed profiles of (a) water vapor pressure (in millibars) and (b) calculated profiles of latent heat flux in a grass canopy at various times during a summer day. [After Ripley and Redmann (1976).]

forestry, and hydrology. More specific applications of agricultural and forest micrometeorology are as follows:

- Determining the mean wind, temperature, humidity, and carbon dioxide profiles in and above a plant cover
- Determining the radiative and other energy fluxes in a plant cover
- Predicting the ground surface and soil temperatures and moisture content
- Estimating evaporation and transpiration losses in plant canopies

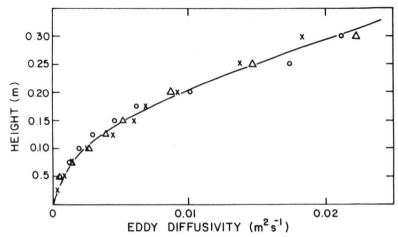

Fig. 15.15 Vertical profiles of eddy diffusivity in a grass canopy calculated by the energy-balance and momentum-balance methods. $\triangle$, Energy balance at 0957; $\bigcirc$, energy balance at 1154; $\times$, momentum balance at 0957–1154. [After Ripley and Redmann (1976).]

• Determining photosynthetic activity and carbon dioxide exchange between plants and the atmosphere
• Protection of vegetation from strong winds and extreme temperature

PROBLEMS AND EXERCISES

1. Explain the concept and physical significance of the zero-plane displacement. How is it related to the height and structure of a plant canopy?
2. (a) Show that the aerodynamic resistance of a plant canopy in a stably stratified surface layer is given by

$$r_{M} = r_{H} = \frac{1}{ku_{*}}\left[\ln\left(\frac{z - d_{0}}{z_{0}}\right) + \beta\left(\frac{z - d_{0}}{L}\right)\right]$$

   (b) Plot r_{M} as a function of the normalized height for the values of $u_{*} = 0.25, 0.50, 0.75$, and 1.0 m sec^{-1} in neutral conditions.
   (c) How does atmospheric stability affect the aerodynamic resistance?
3. (a) Using the Bear–Bouguer law express the net all-wave radiation (R_{N}) as a function of height for a constant leaf area density canopy of height h_{0}.
   (b) For the same canopy, plot $R_{N}(z)/R_{N}(h_{0})$ as a function of z/h_{0} for different values of the extinction coefficient $\alpha = 0.5, 0.75$, and 1.0.

   (c) How is the net longwave radiation expected to vary with height in such a canopy?

4. (a) Using Eq. (15.10) and the mixing-length relation for $\overline{uw}$ with a constant mixing length l_m, derive the exponential wind profile [Eq. (15.11)] for the canopy flow.

   (b) Derive an expression for eddy viscosity K_m for the same flow.

   (c) Plot $U(z)/U(h_0)$ as a function of z/h_0 for different values of the parameter $n = 1, 2, 3,$ and 4.

5. (a) Using the eddy viscosity relation for $\overline{uw}$ with a constant K_m, derive the canopy wind profile Eq. (15.12).

   (b) Compare the wind profiles given by Eqs. (15.11) and (15.12) for $n = m = 3$.

6. (a) Why do the fluxes of momentum, heat, and water vapor generally increase with height in a plant canopy?

   (b) Derive an expression for the momentum flux profile in a uniform plant area density canopy for the given exponential wind profile, Eq. (15.11).

References

Andre, J. C., Moor, G. D., Lacarrere, P., Therry, G., and Vachat, R. (1978). Modeling the 24-hour evolution of the mean and turbulent structures of the planetary boundary layer. *J. Atmos. Sci.* **35**, 1861–1883.

Arya, S. P. S. (1972). The critical condition for the maintenance of turbulence in stratified flows. *Q. J. R. Meteorol. Soc.* **98**, 264–273.

Arya, S. P. S. (1977). Suggested revisions to certain boundary layer parameterization schemes used in atmospheric circulation models. *Mon. Weather Rev.* **105**, 215–227.

Arya, S. P. S. (1978). Comparative effects of stability, baroclinity and the scale-height ratio on drag laws for the atmospheric boundary layer. *J. Atmos. Sci.* **35**, 40–46.

Arya, S. P. S. (1984). Parametric relations for the atmospheric boundary layer. *Boundary-Layer Meteorol.* **30**, 57–73.

Arya, S. P. S. (1986). The schematics of balance of forces in the planetary boundary layer. *J. Clim. Appl. Meteorol.* **24**, 1001–1002.

Arya, S. P. S., and Wyngaard, J. C. (1975). Effect of baroclinity on wind profiles and the geostrophic drag law for the convective planetary boundary layer. *J. Atmos. Sci.* **32**, 767–778.

Arya, S. P. S., Capuano, M. E., and Fagen, L. C. (1987). Some fluid modeling studies of flow and dispersion over two-dimensional low hills. *Atmos. Environ.* **21**, 753–764.

Augstein, E., Schmidt, H., and Ostapoff, F. (1974). The vertical structure of the atmospheric planetary boundary layer in undisturbed trade wind over the Atlantic Ocean. *Boundary-Layer Meteorol.* **6**, 129–150.

Batchelor, G. K. (1970). "An Introduction to Fluid Dynamics." Cambridge Univ. Press, London and New York.

Bradley, E. F. (1968). A micrometeorological study of velocity profiles and surface drag in the region modified by a change in surface roughness. *Q. J. R. Meteorol. Soc.* **94**, 361–379.

Brutsaert, W. H. (1982). "Evaporation into the Atmosphere." Reidel, Boston, Massachusetts.

Businger, J. A. (1973). Turbulent transfer in the atmospheric surface layer. *In* "Workshop on Micrometeorology" (D. A. Haugen, ed.). Amer. Meteorol. Soc., Boston, Massachusetts.

Businger, J. A. (1975). Aerodynamics of vegetated surfaces. *In* "Heat and Mass Transfer in the Biosphere, Part I: Transfer Processes in the Plant Environment" (D. A. deVries and N. H. Afgan, eds.), pp. 139–165. Halstead, New York.

Businger, J. A., Wyngaard, J. C., Izumi, Y., and Bradley, E. F. (1971). Flux-profile relationships in the atmospheric surface layer. *J. Atmos. Sci.* **28**, 181–189.

Byun, D. W. (1987). A two-dimensional mesoscale numerical model of St. Louis Urban mixed layer. Ph.D. dissertation, North Carolina State Univ., Raleigh.

Caughey, S. J., and Palmer, S. G. (1979). Some aspects of turbulence structure through the depth of the convective boundary layer. *Q. J. R. Meteorol. Soc.* **105,** 811–827.

Charnock, H. (1955). Wind stress on a water surface. *Q. J. R. Meteorol. Soc.* **81,** 639–640.

Ching, J. K. S. (1985). Urban-scale variations of turbulence parameters and fluxes. *Boundary-Layer Meteorol.* **33,** 335–361.

Cionco, R. M. (1965). A mathematical model for air flow in a vegetative canopy. *J. Appl. Meteorol.* **4,** 515–522.

Cionco, R. M. (1972). Intensity of turbulence within canopies with simple and complex roughness elements. *Boundary-Layer Meteorol.* **2,** 435–465.

Clarke, R. H., Dyer, A. J., Brook, R. R., Reid, D. G., and Troup, A. J. (1971). The Wangara Experiment: Boundary-layer data. *Tech. Pap. No. 19 CSIRO Div. Meteorol. Phys., Australia.*

Counihan, J. C. (1975). Adiabatic atmospheric boundary layers: A review and analysis of data from the period 1880–1972. *Atmos. Environ.* **9,** 871–905.

Crawford, K. C., and Hudson, H. R. (1973). The diurnal wind variation in the lowest 1500 feet in Central Oklahoma: June 1966–May 1967. *J. Appl. Meteorol.* **12,** 127–132.

Davies, J. A., Robinson, P. J., and Nunez, M. (1970). Radiation measurements over Lake Ontario and the determination of emissivity. First Report, Contract No. H081276, Dept. Geog., McMaster Univ., Hamilton, Ontario.

Deacon, E. L. (1969). Physical processes near the surface of the earth. *In* "World Survey of Climatology, Vol. 2: General Climatology" (H. Flohn, ed.). Elsevier, Amsterdam.

Deardorff, J. W. (1968). Dependence of air-sea transfer coefficients on bulk stability. *J. Geophys. Res.* **73,** 2549–2557.

Deardorff, J. W. (1970). A three-dimensional numerical investigation of the idealized planetary boundary layer. *Geophys. Astrophys. Fluid Dyn.* **1,** 377–410.

Deardorff, J. W. (1972). Parameterization of the planetary boundary layer for use in general circulation models. *Mon. Weather Rev.* **100,** 93–106.

Deardorff, J. W. (1973). Three-dimensional numerical modeling of the planetary boundary layer. *In* "Workshop on Micrometeorology," pp. 271–311. Amer. Meteorol. Soc., Boston, Massachusetts.

Deardorff, J. W. (1978). Observed characteristics of the outer layer. *In* "Short Course on the Planetary Boundary Layer" (A. K. Blackadar, ed.). Amer. Meteorol. Soc., Boston, Massachusetts.

Dutton, J. A. (1976). "The Ceaseless Wind: An Introduction to the Theory of Atmospheric Motion." McGraw-Hill, New York.

Dyer, A. J. (1967). The turbulent transport of heat and water vapor in unstable atmosphere. *Q. J. R. Meteorol. Soc.* **93,** 501–508.

Elliot, W. P. (1958). The growth of the atmospheric internal boundary layer. *Trans. Am. Geophys. Union* **39,** 1048–1054.

Fleagle, R. G., and Businger, J. A. (1980). "An Introduction to Atmospheric Physics," 2nd Ed. Academic Press, New York.

Friehe, C. A., and Schmitt, K. F. (1976). Parameterization of air-sea interface fluxes of sensible heat and moisture by the bulk aerodynamic formulas. *J. Phys. Oceanogr.* **6,** 801–809.

Fritschen, L. J., Driver, C. H., Avery, C., Buffo, J., Edmonds, R., Kinerson, R., and Schiess, P. (1970). Dispersion of air tracers into and within a forested area: 3. *Univ. Wash. Tech. Rep. ECOM-68-68-3,* U.S. Army Electronic Command, Ft. Hauchuca, Arizona.

Gamo, M., Yamamoto, S., Yokoyama, O., and Yoshikado, H. (1983). Structure of the free convective internal boundary layer above the coastal area. *J. Meteorol. Soc. Jpn.* **61,** 110–124.

Garratt, J. R. (1977). Review of drag coefficients over oceans and continents. *Mon. Weather Rev.* **105**, 915–929.

Garratt, J. R., and Brost, R. A. (1981). Radiative cooling effects within and above the nocturnal boundary layer. *J. Atmos. Sci.* **38**, 2730–2746.

Gates, D. M. (1980). "Biophysical Ecology." Springer-Verlag, Berlin and New York.

Geiger, R. (1965). "The Climate Near the Ground." Harvard Univ. Press, Cambridge, Massachusetts.

Gray, W. M., and Mendenhall, B. R. (1973). A statistical analysis of factors influencing the wind veering in the planetary boundary layer. *In* "Climatological Research" (K. Fraedrich, M. Hantel, H. C. Korff, and E. Ruprecht, eds.), pp. 167–194. Dummlers, Bonn.

Halitsky, J. (1968). Gas diffusion near buildings. *In* "Meteorology and Atomic Energy" (D. H. Slade, ed.), pp. 221–255. Tech. Info. Cent., U.S. Depart. Energy, Oak Ridge, Tennessee.

Haltiner, G. J., and Martin, F. L. (1957). "Dynamical and Physical Meteorology-1968." McGraw-Hill, New York.

Hasselmann, K., *et al.* (1973). Measurements of wind wave growth and swell decay during the Joint North Sea Wave Project (JONSWAP). Deutsches Hydrographisches Institut.

Haugen, D. A. (1973). "Workshop on Micrometeorology." Amer. Meteorol. Soc., Boston, Massachusetts.

Hess, S. L. (1959). "Introduction to Theoretical Meteorology." Holt, New York.

Hosker, R. P. (1984). Flow and diffusion near obstacles. *In* "Atmospheric Science and Power Production" (D. Randerson, ed.), pp. 241–326. Tech. Info. Cent., U.S. Dept. Energy, Oak Ridge, Tennessee.

Hunt, J. C. R., and Simpson, J. E. (1982). Atmospheric boundary layer over nonhomogeneous terrain. *In* "Engineering Meteorology" (E. J. Plate, ed.), pp. 269–318. Elsevier, Amsterdam.

Hunt, J. C. R., and Snyder, W. H. (1980). Experiments on stably and neutrally stratified flow over a model three-dimensional hill. *J. Fluid Mech.* **96**, 671–704.

Hutchison, B. A., Matt, D. R., McMillen, R. T., Gross, L. J., Tajchman, S. J., and Norman, J. M. (1986). The architecture of a deciduous forest canopy in Eastern Tennessee, U.S.A. *J. Ecol.* **74**, 635–646.

Inoue, E. (1963). On the turbulent structure of airflow within crop canopies. *J. Meteorol. Soc. Jpn.* **41**, 317–326.

Izumi, Y. (1971). Kansas 1968 Field Program data report. Environ. Res. Pap., No. 379, Air Force Cambridge Research Laboratories, Bedford, Massachusetts.

Izumi, Y., and Barad, M. L. (1963). Wind and temperature variations during development of a low-level jet. *J. Appl. Meteorol.* **2**, 668–673.

Izumi, Y., and Caughey, S. J. (1976). Minnesota 1973 atmospheric boundary layer experiment data report. Environ. Res. Pap., No. 547, Air Force Cambridge Research Laboratories, Bedford, Massachusetts.

Jarvis, P. G., James, G. B., and Landsberg, J. J. (1976). Coniferous forest. *In* "Vegetation and the Atmosphere, Vol. 2, Case Studies" (J. L. Monteith, ed.), pp. 171–240. Academic Press, New York.

Johnson, N. K. (1929). A study of the vertical gradient of temperature in the atmosphere near the ground. *Geophys. Memo.* **46** (Meteorol. Office, London).

Kaimal, J. C., Wyngaard, J. C., Haugen, D. A., Cote, O. R., Izumi, Y., Caughey, S. J., and Readings, C. J. (1976). Turbulence structure in the convective boundary layer. *J. Atmos. Sci.* **33**, 2152–2169.

Kitaigorodski, S. A. (1970). "The Physics of Air–Sea Interaction." Gidromet. Izdatel'stvo, Leningrad (Israel Prog. Sci. Transl., Jerusalem).

Kondo, J. (1976). Heat balance of the East China Sea during the Air Mass Transformation Experiment. *J. Meteorol. Soc. Jpn.* **54**, 382–398.

Kondratyev, K. Y. (1969). "Radiation in the Atmosphere." Academic Press, New York.

Kraus, E. B. (1972). "Atmosphere–Ocean Interaction." Oxford Univ. Press, London and New York.

Kutzbach, J. E. (1961). Investigations of the modifications of wind profiles by artificially controlled surface roughness. Ann. Rep., Contract DA-36-039-SC-80282, University of Wisconsin, pp. 71–114.

Lamb, H. (1932). "Hydrodynamics," 6th Ed. Cambridge Univ. Press, London and New York.

Landsberg, H. E. (1981). "The Urban Climate." Academic Press, New York.

Leavitt, E., and Paulson, C. A. (1975). Statistics of surface layer turbulence over the tropical ocean. *J. Phys. Oceanogr.* **5**, 143–156.

Lenschow, D. H. (1973). Two examples of planetary boundary layer modification over the Great Lakes. *J. Atmos. Sci.* **30**, 568–581.

Lettau, H. H., and Davidson, B. (1957). "Exploring the Atmosphere's First Mile." Pergamon, Oxford.

Lettau, H., Riordan, R., and Kuhn, M. (1977). Air temperature and two-dimensional wind profiles in the lowest 32 meters as a function of bulk stability. *Antarct. Res. Ser.* **25**, 77–91.

Liou, K. N. (1983). "An Introduction to Atmospheric Radiation." Academic Press, New York.

Long, I. F., Monteith, J. L., Penman, H. L., and Szeicz, G. (1964). The plant and its environment. *Meteorol. Rundsch.* **17**, 97–101.

Lowry, W. P. (1967). The climate of cities. *Sci. Am.* **217**, 15–23.

Lowry, W. P. (1970). "Weather and Life: An Introduction to Biometeorology." Academic Press, New York.

McNaughton, K., and Black, T. A. (1973). A study of evapotranspiration from a Douglas fir forest using the energy balance approach. *Water Resour. Res.* **9**, 1579–1590.

Mahrt, L. (1981). The early evening boundary layer transition. *Q. J. R. Meteorol. Soc.* **107**, 329–343.

Mahrt, L., Heald, R. C., Lenschow, D. H., and Stankov, B. B. (1979). An observational study of the structure of the nocturnal boundary layer. *Boundary-Layer Meteorol.* **17**, 247–264.

Mason, P. J., and King, J. C. (1984). Atmospheric boundary layer over a succession of two-dimensional ridges and valleys. *Q. J. R. Meteorol. Soc.* **110**, 821–845.

Mason, P. J., and Sykes, R. I. (1979). Flow over an isolated hill of moderate slope. *Q. J. R. Meteorol. Soc.* **105**, 383–395.

Monin, A. S., and Obukhov, A. M. (1954). Basic turbulent mixing laws in the atmospheric surface layer. *Tr. Geofiz. Inst. Akad. Nauk. SSSR* **24**(151), 163–187.

Monin, A. S., and Yaglom, A. M. (1971). "Statistical Fluid Mechanics: Mechanics of Turbulence," Vol. 1. MIT Press, Cambridge, Massachusetts.

Monji, N. (1972). Budgets of turbulent energy and temperature variance in the transition zone from forced to free convection. Ph.D. thesis, Univ. of Washington, Seattle.

Monteith, J. L. (1973). "Principles of Environmental Physics." Arnold, London.

Monteith, J. L. (1975). "Vegetation and the Atmosphere, Vol. 1: Principles." Academic Press, New York.

Monteith, J. L. (1976). "Vegetation and the Atmosphere, Vol. 2: Case Studies." Academic Press, New York.

Munn, R. E. (1966). "Descriptive Micrometeorology." Academic Press, New York.

Nicholls, S., and LeMone, M. A. (1980). The fair-weather boundary layer in GATE: The relationship of subcloud fluxes and structure to the distribution and enhancement of cumulus clouds. *J. Atmos. Sci.* **37,** 2051–2067.

Obukhov, A. M. (1946). Turbulence in an atmosphere with a non-uniform temperature. *Tr. Inst. Teor. Geofiz., Akad. Nauk. SSSR* **1,** 95–115 (English Transl. in *Boundary-Layer Meteorol.* **2,** 7–29, 1971).

Oke, T. R. (1974). Review of urban climatology 1968–1973. WMO Tech. Note No. 134, World Meteorol. Organ., Geneva.

Oke, T. R. (1978). "Boundary Layer Climates." Halsted, New York.

Oke, T. R. (1987). "Boundary Layer Climates," 2nd Ed., 435 pp. Halsted, New York.

Oke, T. R., and East, C. (1971). The urban boundary layer in Montreal. *Boundary-Layer Meteorol.* **1,** 411–437.

Panofsky, H. A., and Dutton, J. A. (1984). "Atmospheric Turbulence." Wiley (Interscience), New York.

Panofsky, H. A., and Townsend, A. A. (1964). Change of terrain roughness and the wind profile. *Q. J. R. Meteorol. Soc.* **90,** 147–155.

Panofsky, H. A., Tennekes, H., Lenschow, D. H., and Wyngaard, J. C. (1977). The characteristics of turbulent velocity components in the surface layer under convective conditions. *Boundary-Layer Meteorol.* **11,** 355–361.

Paulson, C. A. (1967). Profiles of wind speed, temperature and humidity over the sea. Ph.D. dissertation, Univ. of Washington, Seattle.

Paulson, C. A., Leavitt, E., and Fleagle, R. G. (1972). Air–sea transfer of momentum, heat and water determined from profile measurements during BOMEX. *J. Phys. Oceanogr.* **2,** 487–497.

Pedlosky, J. (1979). "Geophysical Fluid Dynamics." Springer-Verlag, Berlin and New York.

Pendergrass, W., and Arya, S. P. S. (1984). Dispersion in neutral boundary layer over a step change in surface roughness-I. Mean flow and turbulence structure. *Atmos. Environ.* **18,** 1267–1279.

Penman, H. L. (1948). Natural evaporation from open water, bare soil and grass. *Proc. R. Soc. London A* **193,** 120–145.

Pennell, W. T., and LeMone, M. A. (1974). An experimental study of turbulence structure in the fair weather trade wind boundary layer. *J. Atmos. Sci.* **31,** 1308–1323.

Phillips, O. M. (1977). "The Dynamics of the Upper Ocean," 2nd Ed. Cambridge Univ. Press, London and New York.

Plate, E. J. (1971). Aerodynamic characteristics of atmospheric boundary layers. *AEC Crit. Rev. Ser. TID-25465.*

Plate, E. J. (1982). "Engineering Meteorology." Elsevier, Amsterdam.

Pond, S. (1975). The exchanges of momentum, heat and moisture at the ocean–atmosphere interface. "Numerical Models of Ocean Circulation." Nat. Acad. Sci., Washington, D.C.

Prandtl, L. (1905). *Verh. Int. Math. Kongr., 3rd, Heidelburg 1904,* 484–491.

Rao, K. S., Wyngaard, J. C., and Cote, O. R. (1974). The structure of two-dimensional internal boundary layer over a sudden change of surface roughness. *J. Atmos. Sci.* **31,** 738–746.

Raupach, I., and Thom, A. S. (1981). Turbulence in and above plant canopies. *Annu. Rev. Fluid Mech.* **13,** 97–129.

Raynor, G. S., Michael, P., Brown, R. A., and SethuRaman, S. (1975). Studies of atmospheric diffusion from a nearshore oceanic site. *J. Appl. Meteorol.* **14,** 1080–1094.

Raynor, G. S., SethuRaman, S., and Brown, R. M. (1979). Formation and characteristics of coastal internal boundary layers during onshore flows. *Boundary-Layer Meteorol.* **16,** 487–514.

Richardson, L. F. (1920). The supply of energy from and to atmospheric eddies. *Proc. R. Soc. London A* **97,** 354–373.

Rider, N. E., Philip, J. R., and Bradley, E. F. (1963). The horizontal transport of heat and moisture—a micrometeorological study. *Q. J. R. Meteorol. Soc.* **89,** 506–531.

Ripley, E. A., and Redmann, R. E. (1976). Grassland. *In* "Vegetation and the Atmosphere, Vol. 2: Case Studies" (J. L. Monteith, ed.), pp. 351–398. Academic Press, New York.

Rosenberg, N. J., Blad, B. L., and Verma, S. B. (1983). "Microclimate: The Biological Environment," 2nd Ed. Wiley (Interscience), New York.

Ross, J. (1975). Radiative transfer in plant communities. *In* "Vegetation and the Atmosphere, Vol. 1: Principles" (J. L. Monteith, ed.), pp. 13–55. Academic Press, New York.

Royal Aeronautical Society (1972). Characteristics of wind speed in the lower layers of the atmosphere near the ground: Strong winds (neutral atmosphere). Eng. Sci. Data Unit, No. 72026, London.

Schlichting, H. (1960). "Boundary-Layer Theory," 4th Ed. McGraw-Hill, New York.

Sellers, W. D. (1965). "Physical Climatology." Univ. of Chicago Press, Chicago, Illinois.

SethuRaman, S. (1978). Influence of mean wind direction on sea surface wave development. *J. Phys. Oceanogr.* **8,** 926–929.

Sheppard, P. A. (1956). Airflow over mountains. *Q. J. R. Meteorol. Soc.* **82,** 528–529.

Slatyer, R. O., and McIlroy, I. C. (1961). "Practical Micrometeorology." CSIRO Div. Meteorol. Phys., Melbourne.

Smith, S. D. (1980). Wind stress and heat flux over the ocean in gale force winds. *J. Phys. Oceanogr.* **10,** 709–726.

Smith, S. D., and Banke, E. G. (1975). Variation of the sea surface drag coefficient with wind speed. *Q. J. R. Meteorol. Soc.* **101,** 665–673.

Snyder, W. H., Thompson, R. S., Eskridge, R. E., Lawson, R. E., Castro, I. P., Lee, J. T., Hunt, J. C. R., and Ogawa, Y. (1985). The structure of strongly stratified flow over hills: Dividing-streamline concept. *J. Fluid Mech.* **152,** 249–288.

Spangler, T. C., and Dirks, R. A. (1974). Mesoscale variations of the urban mixing height. *Boundary-Layer Meteorol.* **6,** 423–441.

Stanhill, G. (1969). A simple instrument for field measurement of turbulent diffusion flux. *J. Appl. Meteorol.* **8,** 509–513.

Stearns, C. R., and Lettau, H. H. (1963). Report on two wind profile modification experiments in air flow over the ice of Lake Mendota. Annual Report, Contract DA-36-039-AMC-00878, pp. 115–138. University of Wisconsin.

Stunder, M., and SethuRaman, S. (1985). A comparative evaluation of the coastal internal boundary-layer height equations. *Boundary-Layer Meteorol.* **32,** 177–204.

Sutton, O. G. (1953). "Micrometeorology." McGraw-Hill, New York.

Taylor, G. I. (1915). Eddy motion in the atmosphere. *Philos. Trans. R. Soc. London A* **CCXV,** 1–26.

Tennekes, H., and Lumley, J. L. (1972). "A First Course in Turbulence." MIT Press, Cambridge, Massachusetts.

Thom, A. S. (1975). Momentum, mass and heat exchange of plant communities. *In* "Vegetation and the Atmosphere, Vol. 1: Principles" (J. L. Monteith, ed.), pp. 57–109. Academic Press, New York.

Townsend, A. A. (1965). Self-preserving flow inside a turbulent boundary layer. *J. Fluid Mech.* **22,** 773–797.

Townsend, A. A. (1976). "The Structure of Turbulent Shear Flow," 2nd Ed. Cambridge Univ. Press, London and New York.

Uchijima, Z. (1976). Maize and rice. *In* "Vegetation and the Atmosphere, Vol. 2: Case Studies" (J. L. Monteith, ed.), pp. 33–64. Academic Press, New York.

Vehrencamp, J. E. (1953). Experimental investigation of heat transfer at an air–earth interface. *Trans. Am. Geophys. Union* **34,** 22–30.

Venkatram, A. (1977). A model of internal boundary-layer development. *Boundary-Layer Meteorol.* **11,** 419–437.

Vugts, H. F., and Businger, J. A. (1977). Air modification due to a step change in surface temperature. *Boundary-Layer Meteorol.* **11,** 295–305.

Welch, R. M., Cox, S. K., and Davis, J. M. (1980). Solar radiation and clouds. *Meteorol. Monogr.* No. 39.

West, E. S. (1952). A study of annual soil temperature wave. *Aust. J. Sci. Res. Ser. A* **5,** 303–314.

Wilson, J. D., Ward, D. P., Thurtell, G. W., and Kidd, G. E. (1982). Statistics of atmospheric turbulence within and above a corn canopy. *Boundary-Layer Meteorol.* **24,** 495–519.

Woo, H. G. C., Peterka, J. A., and Cermak, J. E. (1977). Wind-tunnel measurements in the wakes of structures. *NASA Contract. Rep.* No. 2806.

Wu, J. (1980). Wind stress coefficients over sea surface near neutral conditions—a revisit. *J. Phys. Oceanogr.* **10,** 727–740.

Wylie, D. P., and Young, J. A. (1979). Boundary-layer observations of warm air modification over Lake Michigan using a tethered balloon. *Boundary-Layer Meteorol.* **17,** 279–291.

Wyngaard, J. C. (1973). On surface-layer turbulence. "Workshop on Micrometeorology," pp. 101–149. Amer. Meteorol. Soc., Boston, Massachusetts.

Wyngaard, J. C. (1984). "Large-Eddy Simulation: Guidelines for Its Application to Planetary Boundary Layer Research." Michaels Communications, Boulder, Colorado.

Wyngaard, J. C., Pennell, W. T., Lenschow, D. H., and LeMone, M. A. (1978). The temperature–humidity covariance budget in the convective boundary layer. *J. Atmos. Sci.* **35,** 47–58.

Index

International Geophysics Series

*Out of print.

Volume 30 ADRIAN E. GILL. Atmosphere–Ocean Dynamics. 1982

Volume 31 PAOLO LANZANO. Deformations of an Elastic Earth. 1982

Volume 32 RONALD T. MERRILL AND MICHAEL W. MCELHINNY. The Earth's Magnetic Field: Its History, Origin and Planetary Perspective. 1983

Volume 33 JOHN S. LEWIS AND RONALD G. PRINN. Planets and Their Atmospheres: Origin and Evolution. 1983

Volume 34 ROLF MEISSNER. The Continental Crust: A Geophysical Approach. 1986

Volume 35 M. U. SAGITOV, B. BODRI, V. S. NAZARENKO, AND KH. G. TADZHIDINOV. Lunar Gravimetry. 1986

Volume 36 JOSEPH W. CHAMBERLAIN AND DONALD M. HUNTEN. Theory of Planetary Atmospheres: An Introduction to Their Physics and Chemistry, Second Edition. 1987

Volume 37 J. A. JACOBS. The Earth's Core, Second Edition. 1987

Volume 38 J. R. APEL. Principles of Ocean Physics. 1987

Volume 39 MARTIN A. UMAN. The Lightning Discharge. 1987

Volume 40 DAVID G. ANDREWS, JAMES R. HOLTON, AND CONWAY B. LEOVY. Middle Atmosphere Dynamics. 1987

Volume 41 PETER WARNECK. Chemistry of the Natural Atmosphere. 1988

Volume 42 S. PAL ARYA. Introduction to Micrometeorology. 1988